OCR

Biology

REVISION GUIDE

David Applin
Ron Pickering

AS

OXFORD

Oxford University Press is a department of the University of Oxford. It furthers the University's objective of excellence in research, scholarship, and education by publishing worldwide in

Oxford New York Auckland Cape Town Dar es Salaam Hong Kong Karachi
Kuala Lumpur Madrid Melbourne Mexico City Nairobi
New Delhi Shanghai Taipei Toronto

With offices in
Argentina Austria Brazil Chile Czech Republic France Greece
Guatamala Hungary Italy Japan South Korea Poland Portugal Singapore
Switzerland Thailand Turkey Ukraine Vietnam

Acc No.

Class No

Date

British Library Cataloguing in Publication Data
Data available

ISBN: 978-0-19913626-1

10 9 8 7 6 5 4 3 2 1

Printed by Bell and Bain Ltd., Glasgow

MIX
Paper from responsible sources
FSC
www.fsc.org FSC® C007785

Acknowledgements:
Authors, editors, co-ordinators and contributors: David Applin, Ron Pickering, Ruth Holmes, Eileen Ramsden, Peter Marshall, Piers Wood, Douglas Griffiths, John Mullins, Sarah Ware.
Diagrams 2.05 meiosis, 2.06 the founder effect, 2.11 mitosis, and 2.17 water transport © David Applin.
Project managed by Elektra Media Ltd. Typeset by Wearset Ltd.

Paper used in the production of this book is a natural, recyclable product made from wood grown in sustainable forests. The manufacturing process conforms to the environmental regulations of the country of origin.

Contents

I hope that this book helps you to do well in your examinations. Although these are an important goal they are not the only reason for studying Biology. I hope also that you will develop a real and lasting interest in the living world which studying Biology will help you to understand. If you enjoy using this book then all the effort of writing it will have been worthwhile.

David Applin

Getting the most out of this Revision Guide

This *Revision Guide* covers the subject content of the specification for the OCR AS Level GCE in Biology H021. It is not intended to replace textbooks and other learning resources but rather to provide succinct coverage of all aspects of the OCR specification, AS Biology.

It is written to be accessible to all students. You will find it useful whether you wish to continue studying Biology after AS or want to achieve as high a grade as possible at AS, as part of a suite of qualifications suitable for continuing studies in other areas.

This book is not a traditional textbook. Where appropriate, content is presented as annotated diagrams, flow charts and graphics integrated with easy-to-read text which nevertheless maintains standards of accuracy and scientific literacy which will enable you to obtain the highest grade possible in your exams. It includes the latest developments in the biological sciences.

Revising successfully

To be successful in AS level Biology you must be able to:
- recall information
- apply your knowledge to new and unfamiliar situations
- carry out precise and accurate experimental work
- interpret and analyse your own experimental data and that of others

Careful revision will enable you to perform at your best in your examinations. Work with determination and tackle the course in small chunks. Make sure that you are active in your revision – just reading information is not enough.

Useful strategies for revision

As well as reading this *Revision Guide*, you might like to use some or all of the following strategies:
- Make your own condensed summary notes.
- Write key definitions onto flash cards.
- Work through facts until you can recall them.
- Ask your friends and family to test your recall.
- Make posters which cover items of the specification for your bedroom walls.
- Carry out exam practice.

Measure your revision in terms of the progress you are making rather than the length of time you have spent working. You will feel much more positive if you are able to say specific things you have achieved at the end of a day's revision rather than thinking 'I spent eight hours inside on a sunny day!' Don't sit for extended periods of time. Plan your day so that you have regular breaks, fresh air, and things to look forward to.

How to improve your recall

Here's a good strategy for recalling information:
- Focus on a small number of facts. Copy out the facts repeatedly in silence for five minutes then turn your piece of paper over and write them from memory.
- If you get any wrong then just write these out for another five minutes.
- Finally test your recall of all the facts.
- Come back to the same facts later in the day and test yourself again.
- Then revisit them the next day and again later in the week.

By carrying out this process the facts will become part of your long-term memory – you will have learnt them!

Once you have built up a solid factual knowledge base you need to test it by completing some past papers for practice. It might be a good idea to tackle several questions on the same topic from a number of papers rather than working through a whole paper at once. This will enable you to identify your weak areas so that you can work on them in more detail.

Finally, remember to complete some mock exam papers under exam conditions.

Answering exam papers

When you look at your exam paper read through all the questions. Identify which are the easiest for you to answer. Start by answering these questions.

Remember to read each question carefully and make sure you are answering the question that is actually set and not the one you would like to be set. Remember to look at the number of marks available for each question and tailor the number of points you make in your answer accordingly. *Do not* write an essay for a question that only attracts one or two marks!

With short-answer questions, look at the amount of space that has been left for the answer. This indicates the length of answer that the examiner anticipates you to give – depending on the size of your handwriting, of course.

Make every effort to answer all the questions. An unanswered question will always score 0! If you can, leave enough time to check through your answers at the end.

Here are some popular words which are often used in exam questions. Make sure you know what the examiner means when each of these words is used.

- **Describe** – Write down all the key points using words and, where appropriate, diagrams. Think about the number of marks that are available when you write your answer.

- **Calculate** – Write down the numerical answer to the question. Remember to include your working and the units.

- **State** – Write down the answer. Remember a short answer rather than a long explanation is required.

- **Suggest** – Use your biological knowledge to answer the question. This term is used when there is more than one possible answer or when the question involves an unfamiliar context.

- **Sketch** – When this word is used a simple freehand answer is acceptable. Remember to make sure that you include any important labels.

- **Define** – Write down what a biological term or statement means.

- **Explain** – Write down a supporting argument using your biological knowledge. Think about the number of marks that are available when you write your answer.

- **List** – Write down a number of points. Think about the number of points required.

- **Discuss** – Write down details of the points in the given topic.

AS GCE Scheme of assessment

The table below shows how the marks are allocated in the OCR AS Biology course.

Unit	Method of assessment	% of AS
AS Unit F211: Cells, Exchange and Transport	1 hour written paper Candidates answer all questions. 60 marks	30%
AS Unit F212: Molecules, Biodiversity, Food and Health	1 hour 45 minutes written paper Candidates answer all questions. 100 marks	50%
AS Unit F213: Practical Skills In Biology 1	Coursework Candidates complete three tasks set by the exam board and marked internally. 40 marks	20%

Assessment objectives

Candidates are expected to demonstrate the following in the context of the biological content of AS Biology:

Knowledge and understanding

- recognise, recall and show understanding of scientific knowledge;
- select, organise and communicate relevant information in a variety of forms.

Application of knowledge and understanding

- analyse and evaluate scientific knowledge and processes;
- apply scientific knowledge and processes to unfamiliar situations including those related to issues;
- assess the validity, reliability and credibility of scientific information.

How science works

- demonstrate and describe ethical, safe and skilful practical techniques and processes, selecting appropriate qualitative and quantitative methods;
- make, record and communicate reliable and valid observations and measurements with appropriate precision and accuracy;
- analyse, interpret, explain and evaluate the methodology, results and impact of their own and others' experimental and investigative activities in a variety of ways.

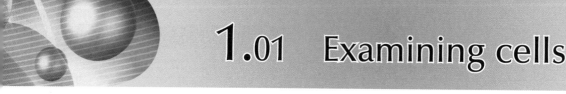

1.01 Examining cells

Seeing cells

Most cells are too small to be seen with the unaided eye. Different types of microscope help us look at cells.

The transmission electron microscope (TEM) reveals the structure of cells in more detail than the optical (light) microscope. The level of detail seen (**resolution of the image**) is a measure of the **resolving power** of the microscope.

- The resolving power of a microscope is its ability to distinguish between structures lying close together.
- Magnifying structures to the limits of a microscope's resolving power reveals more and more detail. But structures lying closer together than the ability of the microscope to resolve them appear as a single structure.
  - As a result, magnifying structures beyond a microscope's resolving power enlarges the image but does not improve clarity of detail.
- The limit of resolution is *one half* the wavelength of the electromagnetic radiation (light or electrons) used to illuminate the specimen under observation.

Magnification is useful when it clarifies the details of a specimen. It is not useful when the limitations of the resolving power of the microscope begin to blur the details. Useful magnification depends on the wavelength of the electromagnetic radiation used to illuminate the specimen.

Property	Light microscope	TEM	SEM
wavelength	400–700 nm	0.005 nm	0.005 nm
resolving power	200 nm	0.5 nm (technical difficulties reduce resolution of the image compared with the theoretical best value)	3–10 nm (less than TEM)
maximum useful magnification	× 2000	× 2 000 000	× 200 000

Electron microscopes

The transmission electron microscope (TEM)

In the TEM a beam of electrons passes *through* the specimen under observation. More electrons pass through some parts of the specimen than others. The electrons then hit a screen coated with phosphorescent material which glows on their impact. The more electron hits, the brighter the glow. The image produced is a highly detailed shadow of the specimen.

The scanning electron microscope (SEM)

In the SEM a beam of electrons *scans* (moves over) the specimen and knocks electrons loose from the specimen's surface. These electrons are captured and a computer processes the information, assembling a detailed three-dimensional image of the specimen's surface features.

The depth of field of view in the SEM is much greater than in the TEM. However its maximum magnification is one order of magnitude (×10) less than the TEM.

Advantages and limitations

High magnification without loss of resolution (the detailed structures of specimens remain clear) is an important advantage of the different types of electron microscope compared with optical microscopes.

However, limitations include

- the expense of buying the instruments, using them and maintaining them
- the difficulty of preparing specimens free of artefacts (features seen in preparations of cells which are not present in living cells)
- their sensitivity to magnetic fields

Qs and As

Q Why is it necessary to have a vacuum inside the body of an electron microscope?

A *Electrons are deflected when they strike molecules of the gases which make up air, making it difficult to focus the electron beam. Producing a vacuum within the microscope solves the problem.*

Staining

Most biological material is transparent. **Staining** is the treatment of the specimen with chemicals such as dyes that improve the contrast between different structures within the cell. Different stains emphasize different chemicals or organelles. For example, in light microscopy:

- Methylene blue stains the nucleus and nucleolus blue.
- Iodine stains starch grains deep blue-purple.
- Acetic orcein stains chromosomes deep red.
- Gram staining stains some bacteria deep blue-black – these bacteria are termed Gram positive. Gram-negative bacteria appear red or pink with Gram staining.

With the TEM, compounds of heavy metals are used to stain the cell as these absorb or scatter the electron beam. These treatments may well change the original tissue, causing artefacts (features visible in the micrograph that were not present in the original tissue).

Magnification

Magnification is the number of times larger an image is than an object. So

$$\text{Magnification} = \frac{\text{size of image}}{\text{size of object}} = \frac{\text{measured length}}{\text{actual length}}$$

- because this is a ratio **there are no units**;
- the same unit must be used for both the size of the image and the size of the object, so some conversion might be necessary, e.g. image measures 70 mm and object measures 3.5 μm.

Convert image size to μm: 70 mm = 70 000 μm

So magnification $= \dfrac{70\,000}{3.5} = 20\,000$ (usually written × 20 000)

If you remember $\text{Mag} = \dfrac{\text{Measured}}{\text{Actual}}$, you can answer any question if you are

given or can measure two of these three quantities.

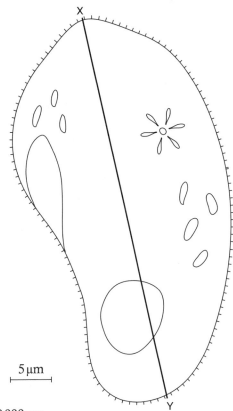

5 μm

1. The scale bar measures 10 mm = 10 000 μm

   The actual length is given as 5 μm

   So magnification $= \dfrac{10\,000}{5} = \times\,2000$

2. To work out the actual size of the specimen, measure X–Y

   Then $2000 = \dfrac{\text{measured X–Y}}{\text{actual X–Y}}$, so

   Actual X–Y $= \dfrac{\text{measured X–Y}}{2000}$

Fact file

SI (Systeme Internationale) units of length are related to each other by 1000 (10^3)

1 m = 1000 mm	1 mm $= = \frac{1}{1000}$ m
1 mm = 1000 μm	1 μm $= = \frac{1}{1000}$ mm
1 μm = 1000 nm	1 nm $= = \frac{1}{1000}$ μm

Questions

1 Explain the difference between the magnifying power and the resolving power of a microscope.

2 Why is the resolving power of a transmission electron microscope greater than that of an optical microscope?

Cell structure

The optical microscope shows that epithelial cells from the intestine are each a unit of **protoplasm** surrounded by a **plasma membrane**. The protoplasm consists of **cytoplasm** in which is embedded a spherical **nucleus**. Notice that the cytoplasm appears speckled.

The resolving power of a transmission electron microscope (TEM) shows the details of **organelles** and other structures not seen even with the best optical microscope.

Organelles are compartments within the cell bounded by membranes which separate the organelles one from another. Compartmentalization within a cell allows activities which would otherwise interfere with one another to take place at the same time.

Other eukaryotic organelles visible under the TEM that are not shown in the diagram below include

- **chloroplasts** in plant cells – containing the pigment chlorophyll (light absorption) and the enzymes necessary for the production of glucose by photosynthesis
- **flagella** and **cilia** – threadlike extensions of the cell having a structure of microtubules. They are involved in cell locomotion.

The role of the cytoskeleton

The cytoplasm contains microfilaments and microtubules, which together make up the cell cytoskeleton. These are threads of the protein actin, situated in bundles just beneath the cell surface. They play a role in the exchange of materials through the cell surface membrane by endo- and exocytosis, and possibly in cell movement. The cytoskeleton also provides mechanical strength to cells.

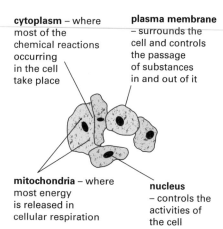

cytoplasm – where most of the chemical reactions occurring in the cell take place

plasma membrane – surrounds the cell and controls the passage of substances in and out of it

mitochondria – where most energy is released in cellular respiration

nucleus – controls the activities of the cell

A representation of epithelial cells as seen through an optical microscope

microvilli – extensions of the plasma membrane increasing the surface area of the cell, maximizing the rate of absorption of substances across the plasma membrane

centrioles – a pair of structures, held at right angles to each other, which act as organisers of the nuclear spindle in preparation for the separation of chromosomes or chromatids during nuclear division

smooth endoplasmic reticulum – a series of flattened sacs and sheets that are the sites of synthesis of steroids and lipids

cell surface membrane – surface of the cell and its contact with the environment. It is differentially permeable and regulates the movement of solutes between the cell and its environment

rough endoplasmic reticulum – so called because of the many ribosomes attached to its surface. This intracellular membrane aids cell compartmentalisation and transports proteins synthesised at the ribosomes towards the Golgi bodies for secretory packaging.

cytoplasm – a solution of different substances through which runs the cytoskeleton, giving support and shape to the cell

nucleus containing chromatin (made of DNA and protein)

glycogen granule

microfilaments

Golgi apparatus – a stack of sac-like structures which package different substances (e.g. carbohydrates and proteins forming glycoproteins). Vesicles bud off, filled with packaged substances. They pass to the plasma membrane where the substances are secreted (released from the cell).

nuclear envelope – double nuclear membrane crossed by a number of nuclear pores. The nuclear envelope is continuous with the endoplasmic reticulum

pore in nuclear envelope

mitochondrion – where energy is released as the result of the reactions of aerobic respiration

lysosomes – sac-like structures which contain high levels of digestive enzymes which would destroy the cell if released into the cytoplasm. The alkaline environment within lysosomes inactivates the enzymes. Lysosomes are abundant in cells like phagocytes.

microtubules

ribosomes – bead-like structures on the rough endoplasmic reticulum and in the cytoplasm. Proteins are made on the ribosomes

nucleolus – darker staining region of the nucleus consisting of proteins and nucleic acids

A representation of an epithelial cell as seen through a TEM

Synthesising proteins

Cell membranes and organelles are involved in the production and secretion of proteins.

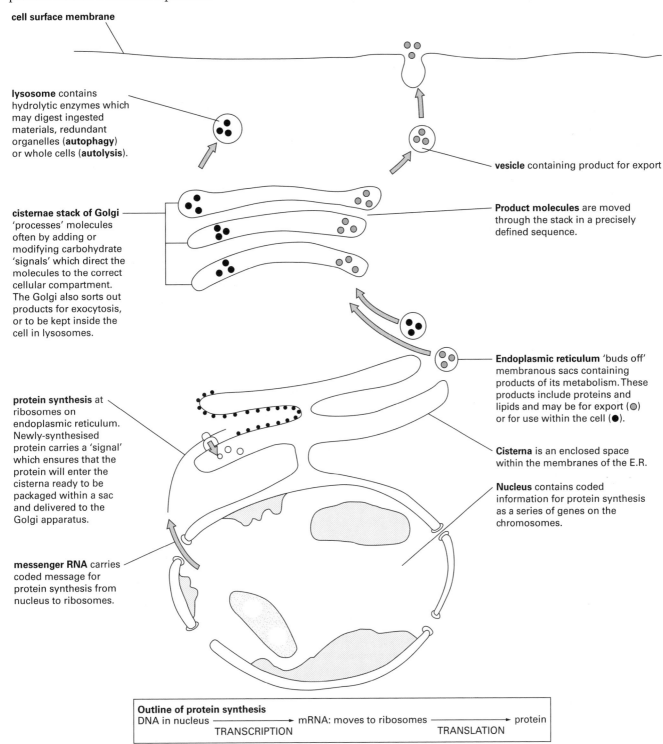

cell surface membrane

lysosome contains hydrolytic enzymes which may digest ingested materials, redundant organelles (**autophagy**) or whole cells (**autolysis**).

vesicle containing product for export

cisternae stack of Golgi 'processes' molecules often by adding or modifying carbohydrate 'signals' which direct the molecules to the correct cellular compartment. The Golgi also sorts out products for exocytosis, or to be kept inside the cell in lysosomes.

Product molecules are moved through the stack in a precisely defined sequence.

Endoplasmic reticulum 'buds off' membranous sacs containing products of its metabolism. These products include proteins and lipids and may be for export (○) or for use within the cell (●).

protein synthesis at ribosomes on endoplasmic reticulum. Newly-synthesised protein carries a 'signal' which ensures that the protein will enter the cisterna ready to be packaged within a sac and delivered to the Golgi apparatus.

Cisterna is an enclosed space within the membranes of the E.R.

Nucleus contains coded information for protein synthesis as a series of genes on the chromosomes.

messenger RNA carries coded message for protein synthesis from nucleus to ribosomes.

Outline of protein synthesis
DNA in nucleus ——————→ mRNA: moves to ribosomes ——————→ protein
                TRANSCRIPTION   TRANSLATION

Questions

1 Explain the importance of compartmentalization in cells.

2 Describe the role of the following in cells:

  a ribosomes

  b mitochondria

  c endoplasmic reticulum

  d Golgi apparatus

Fact file

The term **eukaryotic** refers to cells that have a distinct nucleus. The cells of plants, animals, fungi, and protists (single-celled organisms) are eukaryotic.

Prokaryotic cells do not have membrane-bound organelles. All bacterial cells are prokaryotic cells.

Prokaryotic and eukaryotic cells

A prokaryotic cell as seen under the TEM

*Genetic material is composed of a circle of double-stranded DNA **which is not enclosed within a nuclear membrane**. There are typically about 2000 genes, much less than the number found in a eukaryotic cell.

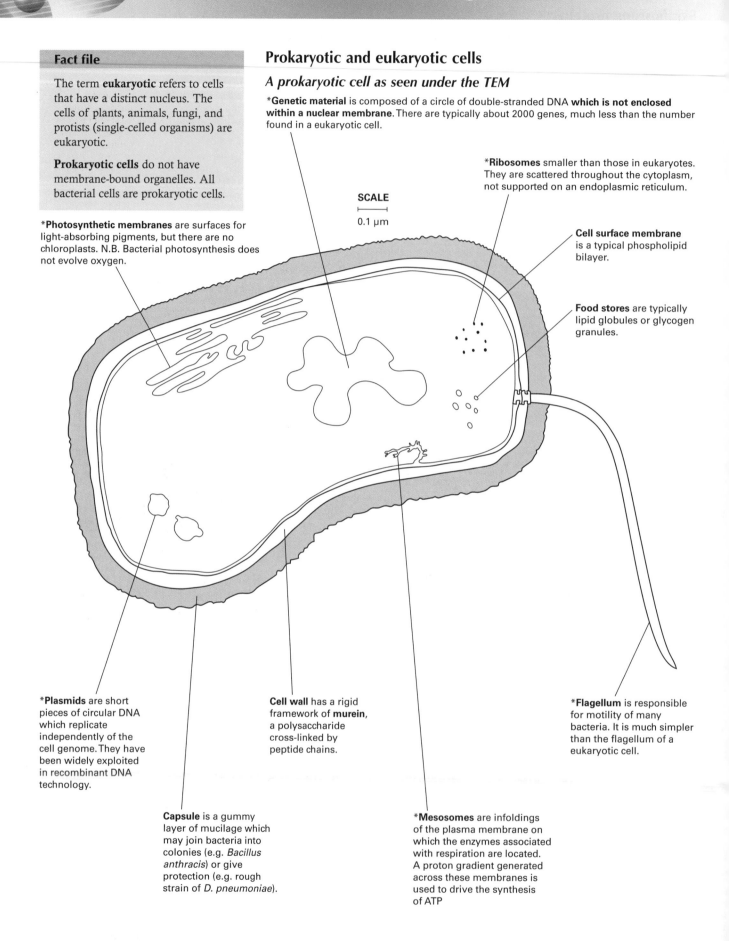

*Ribosomes smaller than those in eukaryotes. They are scattered throughout the cytoplasm, not supported on an endoplasmic reticulum.

Cell surface membrane is a typical phospholipid bilayer.

SCALE
⊢──────⊣
0.1 μm

*Photosynthetic membranes are surfaces for light-absorbing pigments, but there are no chloroplasts. N.B. Bacterial photosynthesis does not evolve oxygen.

Food stores are typically lipid globules or glycogen granules.

*Plasmids are short pieces of circular DNA which replicate independently of the cell genome. They have been widely exploited in recombinant DNA technology.

Cell wall has a rigid framework of **murein**, a polysaccharide cross-linked by peptide chains.

*Flagellum is responsible for motility of many bacteria. It is much simpler than the flagellum of a eukaryotic cell.

Capsule is a gummy layer of mucilage which may join bacteria into colonies (e.g. *Bacillus anthracis*) or give protection (e.g. rough strain of *D. pneumoniae*).

*Mesosomes are infoldings of the plasma membrane on which the enzymes associated with respiration are located. A proton gradient generated across these membranes is used to drive the synthesis of ATP

The cholera bacterium (*Vibrio cholerae*). Important comparisons with eukaryotic cells are marked*

Plant and animal cells

Plant cells share many organelles with animal cells. Compare this diagram of a typical plant cell as seen under the TM with the animal cell on page 10.

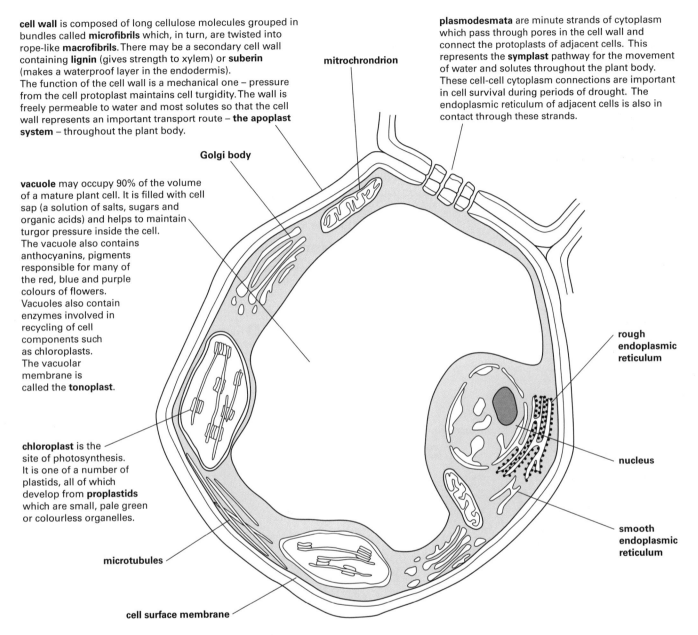

cell wall is composed of long cellulose molecules grouped in bundles called **microfibrils** which, in turn, are twisted into rope-like **macrofibrils**. There may be a secondary cell wall containing **lignin** (gives strength to xylem) or **suberin** (makes a waterproof layer in the endodermis). The function of the cell wall is a mechanical one – pressure from the cell protoplast maintains cell turgidity. The wall is freely permeable to water and most solutes so that the cell wall represents an important transport route – **the apoplast system** – throughout the plant body.

vacuole may occupy 90% of the volume of a mature plant cell. It is filled with cell sap (a solution of salts, sugars and organic acids) and helps to maintain turgor pressure inside the cell. The vacuole also contains anthocyanins, pigments responsible for many of the red, blue and purple colours of flowers. Vacuoles also contain enzymes involved in recycling of cell components such as chloroplasts. The vacuolar membrane is called the **tonoplast**.

chloroplast is the site of photosynthesis. It is one of a number of plastids, all of which develop from **proplastids** which are small, pale green or colourless organelles.

plasmodesmata are minute strands of cytoplasm which pass through pores in the cell wall and connect the protoplasts of adjacent cells. This represents the **symplast** pathway for the movement of water and solutes throughout the plant body. These cell-cell cytoplasm connections are important in cell survival during periods of drought. The endoplasmic reticulum of adjacent cells is also in contact through these strands.

mitrochrondrion

Golgi body

rough endoplasmic reticulum

nucleus

smooth endoplasmic reticulum

microtubules

cell surface membrane

A typical plant cell contains chloroplasts and a permanent vacuole, and is surrounded by a cellulose cell wall

Plant, animal and bacterial cells compared

Feature	Plant	Animal	Bacterium
Cell wall	✓ (cellulose)	✗	✓ (murein)
Nucleus	✓	✓	✗
Plasmids	✗	✗	✓
Mitochondria	✓	✓	✗
Ribosomes	✓	✓	✓ (but small)
Chloroplasts	✓	✗	✗
Permanent vacuole	✓	✗	✗

1.04 Plasma membranes

The roles of plasma membranes

The organelles of eukaryotic cells, as well as the cells themselves, are surrounded by plasma membranes. Prokaryotic cells also have a cell surface membrane. These membranes are important features of the cell:

- They are partially permeable barriers which regulate the movement of solutes between the cell and its environment.
- They separate the cell from its surroundings, allowing control of its internal environment.
- In eukaryotic cells, they also compartmentalise the contents of organelles from the rest of the cell, allowing many activities to take place at the same time without interfering with each other. For example, lysosomes contain harmful digestive enzymes which are separated from the contents of the cell by the plasma membrane.
- Microvilli on the cell surface membrane vastly increase the surface area of some cells, maximising the rate of absorption.

Structure of the plasma membrane

The diagrams with their checklists are your guide to the structure and function of the plasma membrane.

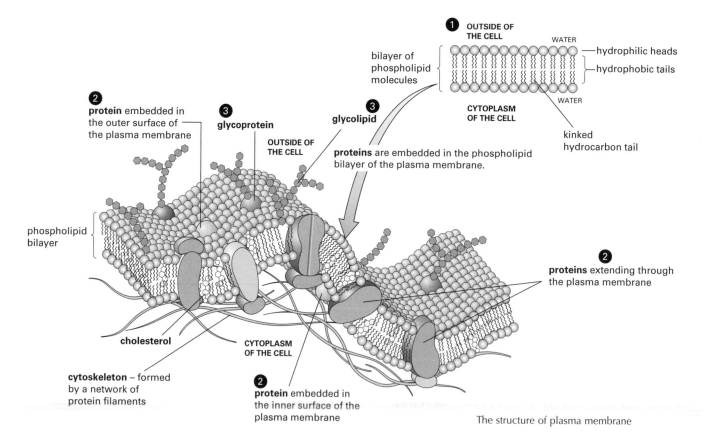

The structure of plasma membrane

Checklist ❶

- **Phospholipids** are an important part of the plasma membrane. Remember that they have hydrophilic heads (phosphate groups) and hydrophobic tails (hydrocarbon chains). Remember also that cell contents are aqueous and cells are bathed in water.

  ⓡ As a result, in water phospholipid molecules spontaneously form a stable two layer framework called a phospholipid **bilayer**. The hydrophobic tails point inwards, shielded from the water. The hydrophilic heads face outwards, forming hydrogen bonds with the water.

- The hydrocarbon tails of many of the phospholipid molecules are 'kinked' because of the presence of double bonds.
  - As a result phospholipids are loosely packed together.
  - As a result the plasma membrane is fluid.
  - As a result proteins and other material can move sideways within the membrane.
- Cholesterol is a lipid. Its molecules, wedged into the bilayer, help to keep the plasma membrane fluid at low temperatures.

Checklist 2

- Different **proteins** are embedded in the phospholipid bilayer.
- Some proteins extend through the membrane; others are localized on one side of the membrane or the other.

The different proteins perform many of the functions of the plasma membrane. Functions include:

- action as enzymes – for example, enzymes on the surfaces of cells lining the intestine catalyse the reactions which digest food
- transport of substances across the membrane
- maintenance of cell shape
- formation of structures which stick cells together
- binding of messenger molecules. The proteins to which messenger molecules bind are called **receptors**. The messenger molecules trigger particular activities in the cell.

Checklist 3

- Sugars bond to proteins and lipids embedded in the outside surface of the plasma membrane.
- A combination of a carbohydrate and a protein forms a **glycoprotein**. A combination of a carbohydrate and a lipid forms a **glycolipid**.
- Glycolipids are receptors for chemicals that are signals between cells. Cells therefore recognize other cells. Glycoproteins work in a similar way.
- Glycoproteins also enable cells of the immune system to recognize foreign cells such as bacteria, which may cause disease.
- Glycolipids and glycoproteins vary from species to species and from one individual to another of the same species. They are identification tags on the surface of cells and unique to the individual.

Temperature changes and the plasma membrane

Cells and their membranes function best at body temperature.

If the temperature rises:

- The fatty acid tails of the phospholipid bilayer become more fluid and allow more movement. This affects the permeability of the cell which may allow harmful substances in, damaging the cell.
- The proteins in the membrane may denature, affecting their function.

If the temperature falls:

- The fatty acid tails of the phospholipids become more rigid which makes the membrane less fluid, impairing the ability of the cell to move and grow.
- Membranes also become less permeable at low temperatures so that vital molecules cannot get into or out of the cell.

Fact file

Membranes are often described as a **fluid mosaic**.

- Their fluidity arises from the loose packing of the molecules of the phospholipid bilayer – this allows sideways movements of proteins and other materials.
- The word 'mosaic' describes the scattering of the different protein molecules embedded within the phospholipid bilayer.

Questions

1 Outline the roles of plasma membranes in eurokaryotic cells.

2 Explain why the plasma membrane of a cell is described as a fluid mosaic.

3 The diagram represents a molecule of phospholipids. Identify the components **x**, **y**, **z**.

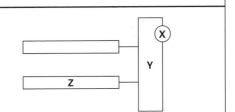

13

Cell signalling

The activities of cells are coordinated throughout the organism. Cells can detect and respond to changes in their immediate surroundings as part of a complex communication system which makes sure that cells perform their functions to the correct extent and at the right time.

Cells communicate with each other:

- by direct contact – for example, by physical contact between the cells, or their cytoplasm is connected by gap junctions
- over large distances – for example, via hormones.

The role of membrane receptors

Receptor proteins in the cell membrane have a specific shape which allows them to bind to other molecules with a complementary shape, such as neurotransmitters, hormones or drugs. The binding can set off a reaction in the cell, for example:

- The neurotransmitter acetylcholine can bind to the postsynaptic membrane and cause an inflow of sodium ions as a nerve impulse is conducted.
- Growth substances may bind and control nuclear division in cells within a tissue.
- The hormone adrenaline can bind to its receptor protein and set off reactions leading to the release of glucose in the cell.

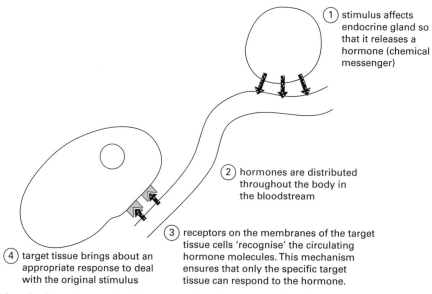

1. stimulus affects endocrine gland so that it releases a hormone (chemical messenger)

2. hormones are distributed throughout the body in the bloodstream

3. receptors on the membranes of the target tissue cells 'recognise' the circulating hormone molecules. This mechanism ensures that only the specific target tissue can respond to the hormone.

4. target tissue brings about an appropriate response to deal with the original stimulus

The role of receptor proteins in hormone action. Hormones produced in one part of the organism are transported throughout the organism and produce a specific response in target cells

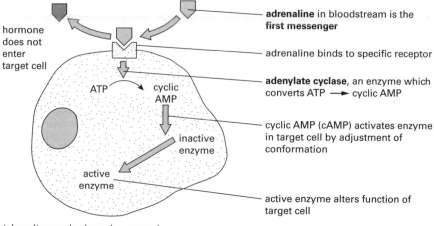

hormone does not enter target cell

adrenaline in bloodstream is the **first messenger**

adrenaline binds to specific receptor

adenylate cyclase, an enzyme which converts ATP ⟶ cyclic AMP

cyclic AMP (cAMP) activates enzyme in target cell by adjustment of conformation

ATP ⟶ cyclic AMP

inactive enzyme

active enzyme

active enzyme alters function of target cell

Adrenaline works through a second messenger

1.06 Diffusion, active transport, and cytosis

Diffusion

Diffusion is the net movement of a substance through a gas or solution from a region where the substance is in high concentration to a region where it is in low concentration. Diffusion continues until the concentration of the substance is the same throughout the gas or solution.

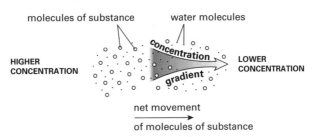

Diffusion

Different factors affect the rate of diffusion:

- The **concentration gradient** of a substance is the difference in concentration between the regions through which the substance is diffusing, divided by the distance.
- The bigger the difference between regions of high concentration and low concentration of a substance, the bigger the concentration gradient.
  - ® As a result the rate of diffusion is maximized.
- The rate of diffusion of a substance decreases in proportion to the square of the distance over which diffusion is taking place (rate ∝ $1/distance^2$).
  - ® As a result diffusion is only effective as a mechanism for the transport of substances over very short distances.
  - ® Diffusion is more effective across thin membranes.
  - ® As a result the size of cells is limited because diffusion is an important mechanism for the transport of substances across membranes within cells.
- A membrane with a larger **surface area** allows more diffusion to take place.

Qs and As

Q Molecules move as a result of their kinetic energy. Their movement is random. In other words there is an equal probability of any one molecule moving in any one of a complete range of directions. Why, therefore, do molecules move down a concentration gradient?

A *If molecules are more highly concentrated in a particular region, then more molecules are more likely to spread to where they are less concentrated than to where they are more concentrated, even though each molecule is moving randomly in any direction. There is, therefore, a net movement of molecules down their concentration gradient: overall more molecules move in a particular direction than in any other direction.*

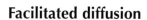

Facilitated diffusion

To enter or leave a cell, most hydrophilic molecules (and ions) diffuse through pores formed by different carrier proteins which cross from one side of the plasma membrane to the other. The process is called **facilitated diffusion**.

The pores of the different carrier proteins are filled with water, making a hydrophilic channel through the hydrophobic region of the phospholipid bilayer. So hydrophilic molecules (and ions) can pass through the channels and therefore through membranes more easily (the word 'facilitated' means 'made easy').

Pores are specific to the substances that pass through them. The rate at which a substance diffuses through the pores of a carrier protein depends on:

- the steepness of its concentration gradient across the membrane – the bigger the gradient, the greater is the rate of diffusion

- the type of carrier protein forming the pore – each type is the right size and structure to allow the passage of a particular substance

- the number of pores in the membrane – the more pores there are, the greater is the rate of diffusion

- whether the pores are open or not

Some pores are open all the time. Others are closed, but they open in the presence of a particular molecule (or ion). Such pores are said to be **gated**. The operation of nerves and muscles depends on gated channels for the movement of sodium ions (Na^+), potassium ions (K^+), and calcium ions (Ca^{2+}) into and out of nerve cells and muscle cells.

As a substance passes through the membrane its carrier protein undergoes a change in tertiary or quaternary structure.

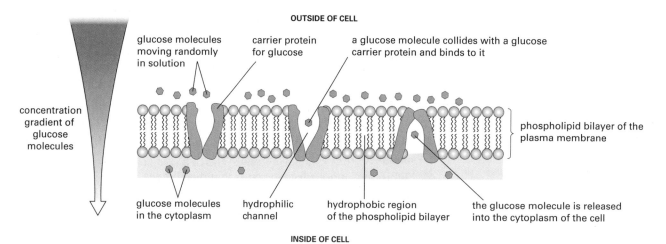

Facilitated diffusion through the plasma membrane. Notice that the binding site for glucose on the carrier protein faces outwards in one state and into the cell in the other state. The attachment of the glucose molecule is responsible for the change in shape and results in the release of glucose into the cytoplasm of the cell.

Active transport

Sometimes molecules or ions move across membranes from where they are in lower concentration to where they are in higher concentration. In other words they move in the reverse direction to diffusion. The process is called **active transport** and allows cells to build up stores of a substance that would otherwise be spread out by diffusion. The storage of glucose by liver cells is an example.

- The process is **active** in the sense that more energy is required to move the molecules or ions against their concentration gradient than down it. This is the energy released by the hydrolysis of ATP which is produced during cellular respiration.

- Diffusion, facilitated diffusion, and osmosis are **passive** processes in the sense that energy provided by the hydrolysis of ATP is not required – the kinetic energy provided as the result of the kinetic motion of their molecules is sufficient to move molecules or ions down their concentration gradient.

Although active transport requires ATP and facilitated diffusion does not, the movement of molecules and ions by both processes is achieved by carrier proteins. Each type of carrier protein is specific for a particular molecule or ion.

The **sodium–potassium pump** is an example of a carrier protein which actively transports sodium (Na^+) ions and potassium (K^+) ions across the plasma membrane of cells.

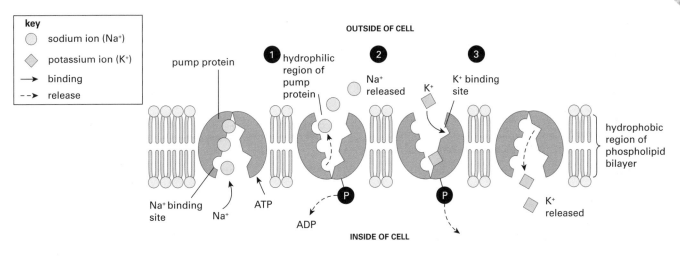

The sodium–potassium pump. Carrier proteins which transport substances in *opposite* directions are called **antiports**.

Checklist for the sodium–potassium pump

1 • The concentration of sodium ions (Na^+) is higher outside the cell than inside.

• The concentration of potassium ions (K^+) is higher inside the cell than outside.

• Three Na^+ and one molecule of ATP bind to the pump protein.

2 • The ATP is hydrolysed and ADP is released. The phosphate group remains bound to the pump protein.

• As a result the shape of the pump protein changes.

• As a result the three Na^+ pass from the cell *against* the concentration gradient of Na^+ and are released.

3 • Two K^+ bind to the pump protein.

• The phosphate group bound to the pump protein is released.

• As a result the structure of the pump protein changes back to its original shape.

• As a result the two K^+ pass into the cell *against* the concentration gradient for K^+ and are released.

Endocytosis and exocytosis

Large molecules such as proteins cross membranes by a process called **cytosis**. Energy (as ATP) is needed, and the process is made possible by the flexibility of the membrane.

• Exocytosis – a vesicle containing the molecules fuses with the inside of the cell surface membrane and the molecule is expelled from the cell.

• Endocytosis – the membrane recognises and binds to a molecule in its environment. The fluid membrane then forms a vesicle (sac) around the molecule, and the sac enters the cell. Phagocytosis (uptake of solids) and pinocytosis (uptake of fluids) are examples of endocytosis.

Summary of diffusion, active transport, and cytosis

Process	Proteins	ATP/energy	Concentration gradient	For example
Diffusion	May be pore proteins	No	Down	Water, Na^+
Facilitated diffusion	Yes	No	Down	Glucose, amino acids in gut
Active transport	Yes	Yes	Up or down (usually up)	Na^+ out of neurones
Endocytosis	Sometimes, to recognise the molecule	Yes	Up or down (usually down)	Milk proteins across baby's gut wall

Question

1 Explain the differences between the processes of diffusion, facilitated diffusion, and active transport.

1.07 Osmosis

What is osmosis?

Osmosis is the net movement of water molecules through a partially permeable membrane from a region where they are in a higher concentration to a region where they are in a lower concentration. The term refers to the diffusion of water molecules and is only used in this context.

Water potential

Water molecules in motion striking a membrane exert pressure called the **water potential**. The higher the concentration of water molecules, the greater is the kinetic energy of the system and the greater is the water potential. Water potential is measured in kilopascals (kPa).

In the diagram, solution A has a higher (less negative) water potential than solution B. There is a water potential gradient between the two solutions across the partially permeable membrane.

Stating the definition of osmosis in terms of water potential:

- Osmosis is the movement of water down a water potential gradient across a partially permeable membrane from a solution of a higher (less negative) water potential to a solution of a lower (more negative) water potential.

The advantages of the water potential nomenclature:

- the movement of water is considered from the 'system's' point of view, rather than from that of the environment;
- comparison between different systems can be made, e.g. between the atmosphere, the air in the spaces of a leaf and the leaf mesophyll cells.

Animal cells and osmosis

Animal cell cytoplasm is a solution of salts and other molecules in water.

The solution surrounding an animal cell can:

- have a higher solute concentration (so a lower water potential) than the cell cytoplasm;
- have the same solute concentration (so the same water potential) as the cell cytoplasm;
- have a lower solute concentration (higher water potential) than the cell cytoplasm.

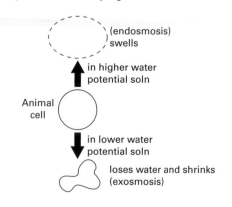

solution A
(dilute solution)

solution B
(concentrated solution)

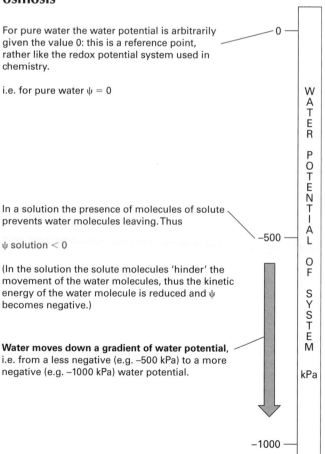

HIGH CONCENTRATION OF WATER — net movement of water molecules → — LOW CONCENTRATION OF WATER

partially permeable membrane

molecules of substance

concentration gradient of water

pore in membrane

water molecules

More water molecules on this side of membrane, so more water molecules pass from left to right.

Fewer water molecules on this side of membrane, so fewer water molecules pass from right to left.

Osmosis

Water potential explains the direction of osmosis

For pure water the water potential is arbitrarily given the value 0: this is a reference point, rather like the redox potential system used in chemistry.

i.e. for pure water $\psi = 0$

In a solution the presence of molecules of solute prevents water molecules leaving. Thus

ψ solution < 0

(In the solution the solute molecules 'hinder' the movement of the water molecules, thus the kinetic energy of the water molecule is reduced and ψ becomes negative.)

Water moves down a gradient of water potential, i.e. from a less negative (e.g. –500 kPa) to a more negative (e.g. –1000 kPa) water potential.

WATER POTENTIAL OF SYSTEM

kPa

0

–500

–1000

Plant cells and osmosis

The presence of solutes in the cell sap makes ψ_s lower (more negative). The ψ for pure water = 0, so a plant cell placed in pure water will experience an inward movement of water, by osmosis, along a water potential gradient.

The entry of water will cause swelling of the protoplast (the cell contents inside the plasma membrane) causing a pressure (**turgor pressure**) to be exerted on the cell wall. Further expansion is resisted by a force **equal, but opposite in sign**, to the turgor pressure. This is called the **wall pressure** or **pressure potential** (ψ_p).

The cellulose cell wall is freely permeable to water and to solutes.

The plasma membrane is freely permeable to water but of limited permeability to solutes.

The pressure potential tends to resist the entry of water or to force water out of the cell.

$$\psi_{Cell} \quad = \quad \psi_s \quad + \quad \psi_p$$

Water potential of cell, i.e. the tendency of water to leave the cell: sometimes written as ψ_w, ψ_o or simply as ψ.

Effect of solute concentration on water potential – this is always negative in value.

Effect of wall pressure and turgor pressure on water potential – it represents the tendency for water to be forced out of the cell, and is either zero or positive in value.

Water relationships of plant cells

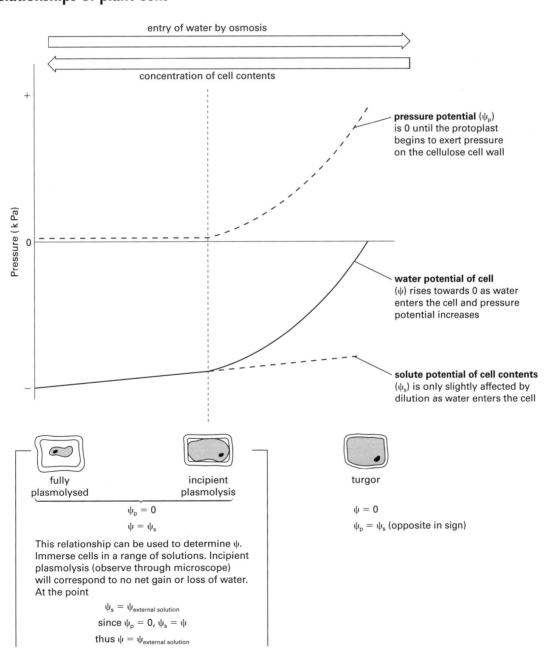

entry of water by osmosis

concentration of cell contents

Pressure (kPa)

pressure potential (ψ_p) is 0 until the protoplast begins to exert pressure on the cellulose cell wall

water potential of cell (ψ) rises towards 0 as water enters the cell and pressure potential increases

solute potential of cell contents (ψ_s) is only slightly affected by dilution as water enters the cell

fully plasmolysed

incipient plasmolysis

turgor

$\psi_p = 0$

$\psi = \psi_s$

$\psi = 0$

$\psi_p = \psi_s$ (opposite in sign)

This relationship can be used to determine ψ. Immerse cells in a range of solutions. Incipient plasmolysis (observe through microscope) will correspond to no net gain or loss of water. At the point

$$\psi_s = \psi_{external\ solution}$$

since $\psi_p = 0$, $\psi_s = \psi$

thus $\psi = \psi_{external\ solution}$

1.08 The cell cycle and mitosis

The cell cycle

New cells are formed from existing cells.

- The cells that give rise to new cells are called **parent** cells.
- The new cells are the **daughter** cells formed when the parent cell divides.

The **cell cycle** describes the process. You can see that nuclear division – mitosis – takes a small fraction of the cell cycle. During the rest of the cycle, genetic information is copied and checked.

key

G_1 – the cell grows and most of its organelles are replicated

S – replication of DNA. Each length of DNA forms a pair of strands. The strands of a pair are identical, each forming a **chromatid** which is the **sister** of its partner. Sister chromatids are joined at the **centromere**.

G_2 – replication of the centrioles

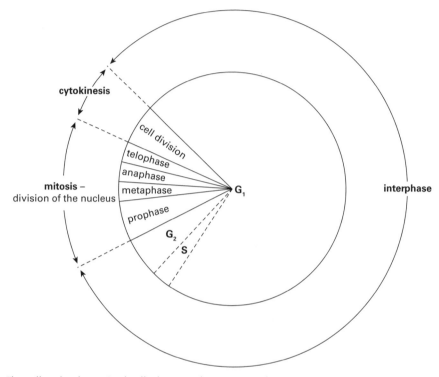

The cell cycle of an animal cell takes 8–24 hours to complete.

Interphase

The genetic material in the cell's nucleus appears under the optical microscope as diffuse **chromatin**. The different phases of interphase are called G_1, **S**, and G_2.

Mitosis

Mitosis follows interphase. The replicated DNA (in the form of chromatids) of the parent cell appears as distinct chromosomes under the optical microscope.

The diagram of mitosis (on the next page) in an animal cell illustrates just four chromosomes as paired chromatids.

- The movements of the chromatids during mitosis in a plant cell are identical, except that spindle formation takes place in the *absence* of centrioles.
- Remember that a cell inherits **two** sets of chromosomes, which is why the parent cell and its daughter cells are said to be **diploid**. The symbol $2n$ represents the diploid state where n = the number in a set of chromosomes ($2n = 4$ in the parent cell and daughter cells of the diagram).

The significance of mitosis

Mitosis increases cell numbers in an organism. The growth and repair of tissues depends on mitosis, as does asexual reproduction in plants and animals.

Fact file

Diploid cells have two sets of chromosomes, one that originated from the male parent, and one from the female parent. Each chromosome has a corresponding matching paired chromosome that is the same shape and carries genes for the same characteristics. These matching chromosomes are called **homologous chromosomes**.

In humans, there are 23 homologous pairs of chromosomes in a diploid cell.

The **diploid** number $2n$ of chromosomes of the parent cell is **four**.

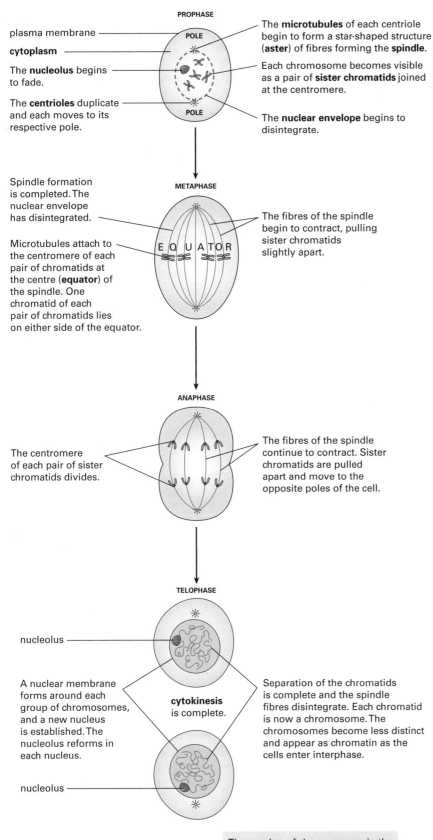

PROPHASE

plasma membrane

cytoplasm

The **nucleolus** begins to fade.

The **centrioles** duplicate and each moves to its respective pole.

POLE

POLE

The **microtubules** of each centriole begin to form a star-shaped structure (**aster**) of fibres forming the **spindle**.

Each chromosome becomes visible as a pair of **sister chromatids** joined at the centromere.

The **nuclear envelope** begins to disintegrate.

METAPHASE

Spindle formation is completed. The nuclear envelope has disintegrated.

Microtubules attach to the centromere of each pair of chromatids at the centre (**equator**) of the spindle. One chromatid of each pair of chromatids lies on either side of the equator.

E Q U A T O R

The fibres of the spindle begin to contract, pulling sister chromatids slightly apart.

ANAPHASE

The centromere of each pair of sister chromatids divides.

The fibres of the spindle continue to contract. Sister chromatids are pulled apart and move to the opposite poles of the cell.

TELOPHASE

nucleolus

A nuclear membrane forms around each group of chromosomes, and a new nucleus is established. The nucleolus reforms in each nucleus.

nucleolus

cytokinesis is complete.

Separation of the chromatids is complete and the spindle fibres disintegrate. Each chromatid is now a chromosome. The chromosomes become less distinct and appear as chromatin as the cells enter interphase.

The number of chromosomes in the nucleus of each daughter cell is **four**: the **diploid** number $2n$.

Mitosis in an animal cell. The stages of mitosis are **prophase**, **metaphase**, **anaphase**, and **telophase**. The memory aid **ProMAT** will help you to remember the sequence.

Fact file

Colchicine is a chemical extracted from crocus corms. Metaphase can be seen more clearly if colchicine is added to a culture of cells undergoing mitosis. Spindle formation is inhibited preventing the separation of chromatids during anaphase.

Feulgen's solution stains DNA dark purple. Staining mitotic cells with Feulgen's solution makes it easier to visualize their chromosomes through the optical microscope.

Cytokinesis

Cytokinesis follows division of the nucleus at the end of telophase. Organelles such as mitochondria and chloroplasts are evenly distributed between the poles of the parent cells as its cytoplasm divides.

- In animal cells filaments of the protein **actin** attach to the inner surface of the plasma membrane in the region of the equator of the parent cells. The filaments contract, pulling the plasma membrane inwards. A 'waist' or **division furrow** forms which deepens, eventually splitting the parent cell into two daughter cells.

- In plant cells the Golgi apparatus forms vesicles containing carbohydrates (e.g. cellulose) in the middle of the parent cell. The vesicles fuse forming a **cell plate** which extends outwards until it meets the plasma membrane. A new cell wall forms, splitting the parent cell into two daughter cells.

Asexual reproduction in yeast

Yeast cells reproduce by asexual reproduction, by the process of mitosis. A bud forms on the parent cell, grows, and then separates from the parent to form a new individual that is genetically identical to the parent.

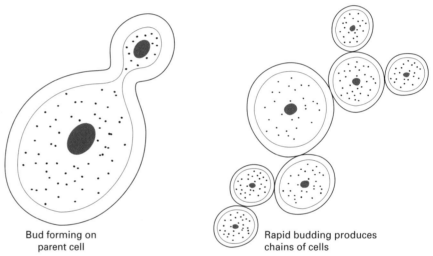

Bud forming on
parent cell

Rapid budding produces
chains of cells

Budding in yeast – asexual reproduction produces genetically identical cells by the process of mitosis.

Question

1 The diagram illustrates the change in DNA content during the cell cycle.

   a Calculate the percentage of the cell cycle time spent in G_1.

   b At which point does chromosome replication begin? Explain your answer.

   c At which point does cytokinesis begin? Explain your answer.

   d At which point does mitosis begin? Explain your answer.

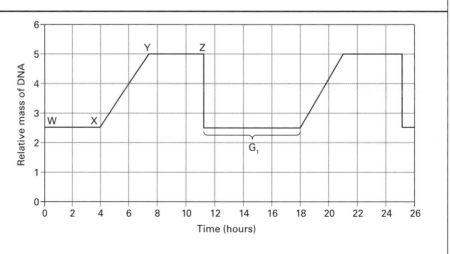

1.09 Meiosis

Cell division by meiosis gives rises to gametes (sex cells). It progresses through the same phases as mitosis (with some differences), but the phases occur twice over.

- The first meiotic division is a **reduction** division which results in two daughter cells, each with half the number of chromosomes of the nucleus of the parent cell. The cells are **haploid**.
- The second meiotic division is a mitosis during which the two haploid daughter cells (resulting from the first meiotic division) divide.

Daughter cells which are haploid rather than diploid are an important difference which distinguishes meiosis from mitosis. The daughter cells are not genetically identical with either the parent cell or each other – they are genetically unique.

The importance of meiosis

- Daughter cells each receive a half (**haploid** or n) set of chromosomes from the parent cell.
    - As a result, during **fertilization** (when sperm and egg join together) the chromosomes of each cell combine.
    - As a result, the **zygote** (fertilized egg) receives a full (**diploid** or $2n$) set of chromosomes, but inherits a new combination of the genes carried on the chromosomes (50:50) from the parents.
    - As a result, the new individual inherits characteristics from both parents, not just from one parent as in asexual reproduction.

What is the result of meiosis?

- The diploid number of chromosomes has been halved.
- Genetic material is exchanged between homologous chromosomes (as a result of crossing over).
- Chromosomes separate randomly.
    - As a result alleles (genes) are separated randomly – a process called **independent assortment**.
    - As a result male (paternal) and female (maternal) chromosomes are distributed randomly among the daughter cells.

Remember that during cell division the chromosomes in the nucleus of the **parent** cell pass to the new **daughter** cells. 'Daughter' does not mean that the cells are female. It means that they are the new cells formed as a result of cell division.

Remember that, strictly, meiosis (and mitosis) refers to the processes which lead to the division of the nucleus of the parent cell. **Cytokinesis** which follows meiosis (and mitosis) describes division of the cell itself.

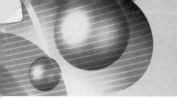
Differentiation

The cells of an embryo at an early stage all look alike. They are **undifferentiated**. As the embryo develops, it forms different tissues for different functions. Each cell in an early embryo has the capacity to develop into any of the types of cells needed to make up these tissues. The cells are called **embryonic stem cells**.

- The term **differentiation** refers to the process which enables undifferentiated cells to develop into particular types of cell.
- Following differentiation, the cells are said to be **differentiated**.

Differentiated cells are **specialized**, enabling them to carry out particular functions – for example producing mucus or conducting nerve impulses.

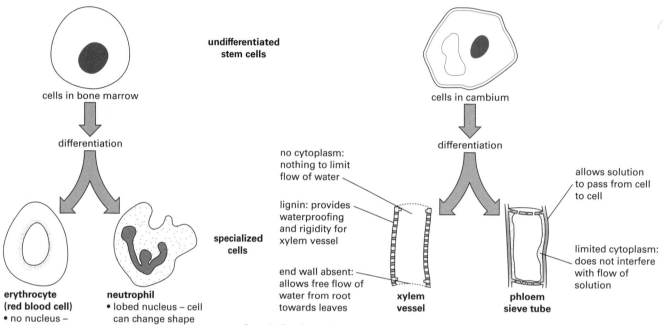

- Specialization adapts cells, enabling them to carry out different tasks.

Differentiation is an exact sequence of events during embryonic development. It lays down the features of the embryo in the right place at the right time.

- The process is controlled by **developmental genes**.
- These genes switch their activity 'on' and 'off' in the correct order to ensure the proper development of an embryo.

Cells are specialized

The human body is made up of more than 200 different types of cell. Each type of cell is adapted enabling it to carry out a particular task (the cell's function).

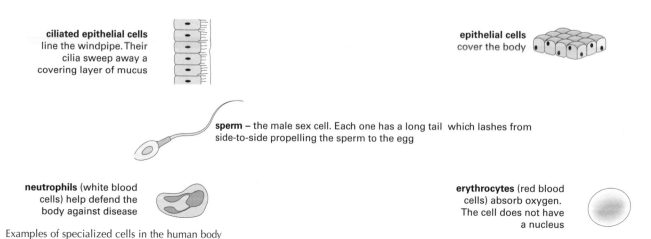

Examples of specialized cells in the human body

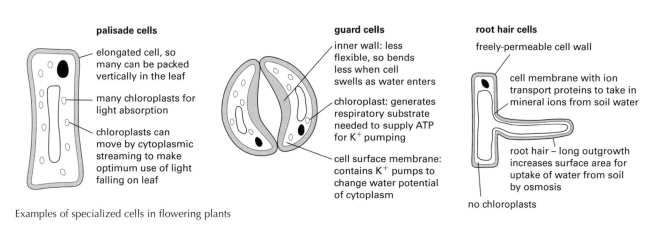

palisade cells

elongated cell, so many can be packed vertically in the leaf

many chloroplasts for light absorption

chloroplasts can move by cytoplasmic streaming to make optimum use of light falling on leaf

guard cells

inner wall: less flexible, so bends less when cell swells as water enters

chloroplast: generates respiratory substrate needed to supply ATP for K$^+$ pumping

cell surface membrane: contains K$^+$ pumps to change water potential of cytoplasm

root hair cells

freely-permeable cell wall

cell membrane with ion transport proteins to take in mineral ions from soil water

root hair – long outgrowth increases surface area for uptake of water from soil by osmosis

no chloroplasts

Examples of specialized cells in flowering plants

Tissues

A **tissue** is a group of cells of the same type. Different types of tissue include connective, muscle, nerve, and epithelial tissue.

Epithelia are tissues which cover the inside and outside of body surfaces. Each type has a particular function:

- **Protection** of internal organs from damage – for example, skin cells are thickened with the protein keratin which helps the cells to resist abrasion.
- **Diffusion** of substances across the surface of the epithelium – for example, the wall of the alveolus is a simple flattened epithelium one cell thick. Oxygen and carbon dioxide easily diffuse across its surface.

- **Absorption** of materials – for example, the free surface of the cells covering the villi of the small intestine are folded into microvilli. The increase in surface area maximizes the rate of absorption of digested food.
- **Secretion** of substances onto the surface of the epithelium – for example, **goblet cells** secrete mucus onto the surface of the epithelium lining the tubes of the trachea and bronchi through which air passes to and from the lungs.

The appearance of epithelial cells seen with the optical microscope is often a clue to their function.

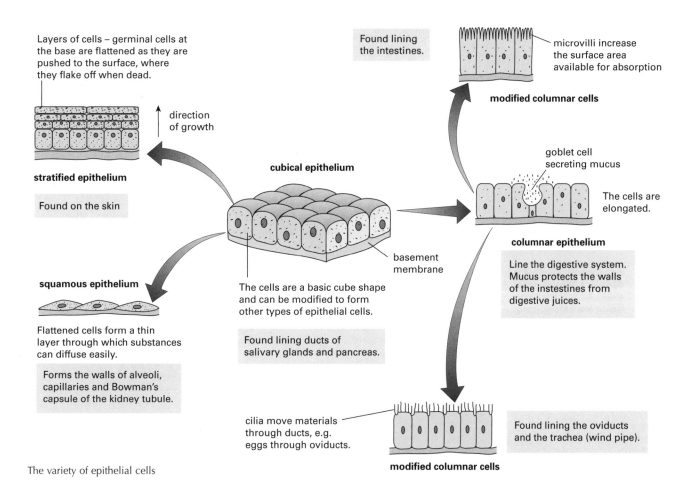

Layers of cells – germinal cells at the base are flattened as they are pushed to the surface, where they flake off when dead.

direction of growth

stratified epithelium

Found on the skin

squamous epithelium

Flattened cells form a thin layer through which substances can diffuse easily.

Forms the walls of alveoli, capillaries and Bowman's capsule of the kidney tubule.

cubical epithelium

basement membrane

The cells are a basic cube shape and can be modified to form other types of epithelial cells.

Found lining ducts of salivary glands and pancreas.

Found lining the intestines.

microvilli increase the surface area available for absorption

modified columnar cells

goblet cell secreting mucus

The cells are elongated.

columnar epithelium

Line the digestive system. Mucus protects the walls of the instestines from digestive juices.

cilia move materials through ducts, e.g. eggs through oviducts.

modified columnar cells

Found lining the oviducts and the trachea (wind pipe).

The variety of epithelial cells

Fact file

The cells, tissues, and organs within an organ system all cooperate – they work in a coordinated way to bring about the function of the organ system.

Questions

1 Briefly explain the role of genes in the development of the embryo.

2 Why is the appearance of epithelial cells often a clue to their function?

3 Briefly explain the difference between a tissue, organ, and organ system.

Organs and organ systems

Different tissues working together make up an **organ**. Different organs combine to make an **organ system**.

- The skin is the largest organ in the human body. Other organs include the liver, kidney, heart, stomach, lungs, brain, ovary and many more.
- Organ systems in the human body include the lymphatic, respiratory, digestive, urinary, reproductive, muscular, skeletal, nervous, endocrine, integumentary (skin, hair, etc.) and circulatory systems.

Plants also have tissues, organs and organ systems. The diagram shows their arrangement in a leaf.

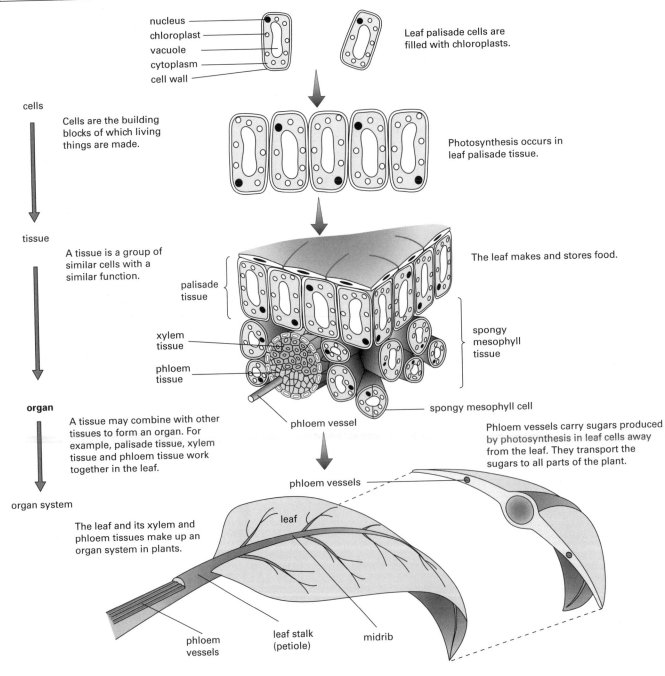

nucleus
chloroplast
vacuole
cytoplasm
cell wall

Leaf palisade cells are filled with chloroplasts.

cells
Cells are the building blocks of which living things are made.

Photosynthesis occurs in leaf palisade tissue.

tissue
A tissue is a group of similar cells with a similar function.

The leaf makes and stores food.

palisade tissue

xylem tissue

phloem tissue

spongy mesophyll tissue

spongy mesophyll cell

phloem vessel

organ
A tissue may combine with other tissues to form an organ. For example, palisade tissue, xylem tissue and phloem tissue work together in the leaf.

Phloem vessels carry sugars produced by photosynthesis in leaf cells away from the leaf. They transport the sugars to all parts of the plant.

phloem vessels

organ system
The leaf and its xylem and phloem tissues make up an organ system in plants.

leaf

phloem vessels

leaf stalk (petiole)

midrib

Exchanging materials across a surface

All cells (tissues, organs, organisms) exchange gases, food and other materials with their environment. These exchanges occur across the surfaces of epithelia. The larger the surface, the more material can be exchanged.

Surface area to volume ratio

The diagram and table show calculations of surface area (SA) and volume (V) for three cubes of different sizes. (Remember that a cube has six faces.)

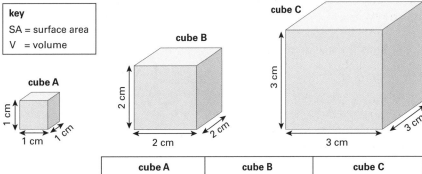

key

SA = surface area
V = volume

	cube A	cube B	cube C
SA of one face	$1 \times 1 = 1$ cm^2	$2 \times 2 = 4$ cm^2	$3 \times 3 = 9$ cm^2
SA of cube	1 cm$^2 \times 6 = 6$ cm^2	4 cm$^2 \times 6 = 24$ cm^2	9 cm$^2 \times 6 = 54$ cm^2
V of cube	$1 \times 1 \times 1 = 1$ cm^3	$2 \times 2 \times 2 = 8$ cm^3	$3 \times 3 \times 3 = 27$ cm^3
$\dfrac{SA}{V}$	$\dfrac{6}{1} = 6$	$\dfrac{24}{8} = 3$	$\dfrac{54}{27} = 2$

Notice:
- The SA/V of cube B is half that of cube A.
- The SA/V of cube C is two-thirds that of cube B and one-third that of cube A.

Remember:
- The *larger* the cube becomes, the *smaller* its SA/V because SA increases more slowly than V.
- Surface area increases with the **square** (power^2) of the side
- Volume increase with the **cube** (power^3) of the side.

Cells (tissues, organs, organisms) are not usually cubical, but the calculations apply to any shape. For example, as a cell grows it
- takes in more nutrients and gases
- produces more waste substances

After the cell reaches a certain size, its surface area becomes too small to take in enough of the substances it needs and remove enough of the wastes it produces.

At this point the cell divides into two smaller daughter cells whose surface area to volume ratio is greater than that of the parent cell, enabling enough
- food and gases to pass into the cells
- wastes to pass out of the cells

Qs and As

Q Why does a small mammal such as a vole (a rodent about 5 cm long) eat its own weight of food each day?

A *A vole's body has a large surface area relative to its volume. Much heat therefore is lost from its body through the skin to the environment. So, it needs lots of food as a source of energy to help maintain a constant body temperature.*

Organ systems specialized for exchanging materials

All organisms exchange gases, food, and other materials between themselves and the environment. The exchanges take place across body surfaces.

Different adaptations increase surface area in multicellular organisms, maximizing the exchange of materials between an organism and its surroundings.

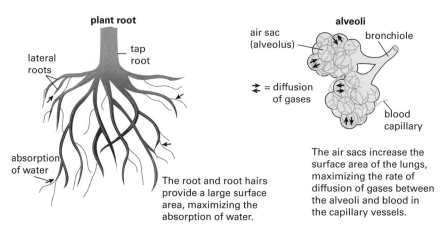

Multicellular organisms have organ systems adapted to increase their surface area to volume ratios

Exchange surfaces

Diffusion across membranes

All organisms exchange oxygen and carbon dioxide between themselves and their environment. The gases are exchanged by diffusion across surface membranes. The membranes have properties that help to maximize the rate of diffusion. They

* are thin – the thinner the membrane, the more quickly gas molecules pass through it
* have a large surface area – the larger the surface area of a membrane, the more molecules of gas collide with it and pass through
* are permeable to the gases

Also, the rate of diffusion of gases across a membrane increases the steeper their respective concentration gradients are on either side of the membrane.

Gas exchange – exchanging oxygen and carbon dioxide with the environment

The chemical reactions of **cellular respiration** release energy which powers the activities of organisms.

* The reactions of **aerobic respiration** (a form of cellular respiration) require oxygen.
    * As a result air containing oxygen passes into organisms which live on land, or oxygen in solution passes into organisms which live in water.
* Carbon dioxide is a waste product of aerobic respiration.
    * As a result carbon dioxide passes to the air or water from organisms.
* In plants the reactions of photosynthesis produce oxygen.
* If the rate of photosynthesis is greater than that of aerobic respiration (for example in bright light) then more oxygen will be produced than the plant requires for aerobic respiration.
    * As a result oxygen passes from the plant to the environment.
* If the rate of photosynthesis is less than that of respiration (for example in dim light) then less oxygen will be produced than the plant requires for aerobic respiration.
    * As a result oxygen passes to the plant from the environment.

Why are exchange surfaces moist?

Carbon dioxide and oxygen would diffuse more quickly as gases than when dissolved in solution in water. However, it is important that exchange surfaces are kept moist to prevent desiccation of the cells and hence destruction of the exchange surface.

Questions

1 Explain why the surface area to volume ratio (SA/V) of large organisms is less than that of small organisms.

2 Briefly explain why an increase in size might be a stimulus for a cell to divide.

1.12 Gas exchange in the lungs

The diagram helps you to locate and identify the important structures of the human gas exchange system.

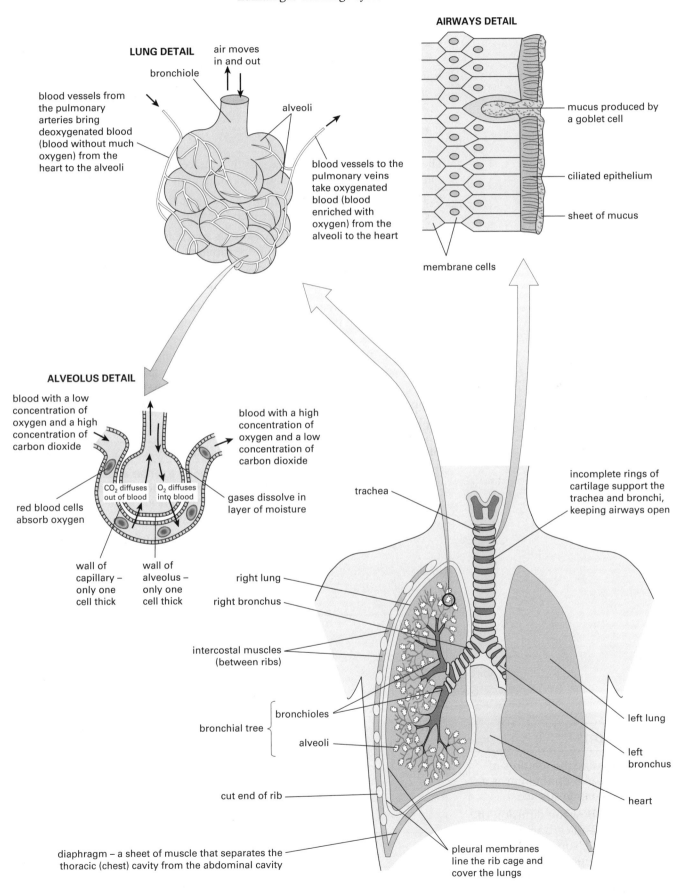

AIRWAYS DETAIL

mucus produced by a goblet cell

ciliated epithelium

sheet of mucus

membrane cells

LUNG DETAIL

bronchiole

air moves in and out

alveoli

blood vessels from the pulmonary arteries bring deoxygenated blood (blood without much oxygen) from the heart to the alveoli

blood vessels to the pulmonary veins take oxygenated blood (blood enriched with oxygen) from the alveoli to the heart

ALVEOLUS DETAIL

blood with a low concentration of oxygen and a high concentration of carbon dioxide

blood with a high concentration of oxygen and a low concentration of carbon dioxide

CO_2 diffuses out of blood

O_2 diffuses into blood

gases dissolve in layer of moisture

red blood cells absorb oxygen

wall of capillary – only one cell thick

wall of alveolus – only one cell thick

trachea

incomplete rings of cartilage support the trachea and bronchi, keeping airways open

right lung

right bronchus

intercostal muscles (between ribs)

bronchial tree

bronchioles

alveoli

left lung

left bronchus

heart

cut end of rib

diaphragm – a sheet of muscle that separates the thoracic (chest) cavity from the abdominal cavity

pleural membranes line the rib cage and cover the lungs

In the diagram of the human gas exchange system, notice that

- the **trachea** (windpipe) branches into two bronchi
- each bronchus branches into many **bronchioles**
- each bronchiole ends in a cluster of **alveoli** (air sacs)
- the alveoli make up the honeycomb lung tissue

Notice also

- A network of capillary vessels supplies blood to and carries blood from the alveoli.
- The walls of the alveoli and capillary vessels are each one cell thick.
  - As a result a surface only two cells thick separates the air in the alveolus and the blood in the capillary blood vessel.
  - As a result the diffusion of gases (oxygen and carbon dioxide) across the surface is rapid.

The rate of diffusion of gases across the surface of the alveoli depends on concentration gradients. Inhalation (breathing in) draws air into the alveoli.

- The concentration of oxygen in inhaled air in the alveoli is greater than the concentration of oxygen in the blood supplied to the alveoli.
  - As a result oxygen diffuses down its concentration gradient from the air in the alveoli to the blood in the capillary vessels supplying the alveoli.
- The concentration of carbon dioxide in the blood supplied to the alveoli is greater than that in the inhaled air in the alveoli.
  - As a result carbon dioxide diffuses down its concentration gradient from the blood in the capillary vessels supplying the alveoli to the air in the alveoli.

Exhalation (breathing out) carries air out of the lungs. The table shows the percentage change in oxygen and carbon dioxide of inhaled and exhaled air.

Adaptations of the mammalian lung to efficient gaseous exchange

Structure	Adaptations and functions
Trachea and bronchi	Provide pathway for air to enter and leave the lungs. Ventilation maintains the concentration gradient for oxygen and carbon dioxide in the alveoli, maximising the rate of diffusion.
	Mucus-secreting cells (goblet cells) and ciliated epithelium clean the incoming air.
	Cartilage rings support the airways and prevent them collapsing as air pressure in the lungs changes during ventilation.
Bronchioles	Finely branching tubes that lead to the alveoli. Terminal bronchioles have no cartilage rings.
	Smooth muscle in the bronchioles allows their diameter to be controlled, e.g. adrenaline relaxes the bronchioles and allows better air flow during exercise.
Alveoli	Very numerous providing a big surface area for diffusion. Walls are one cell thick. In close contact with capillaries, also with walls one cell thick, providing short distance for diffusion.
	Elastic fibres allow alveoli to recoil (spring back) after expiration and allow them to extend during inspiration.

Questions

1 Summarize the features of the alveolar epithelium as a gas exchange surface.

2 Explain how the diffusion of oxygen and carbon dioxide across the surface of the alveoli depends on their respective concentration gradients.

Breathing in and breathing out

The cage around the **thoracic** (chest) **cavity** formed by the ribs and diaphragm is elastic. As it moves the pressure of air in the lungs changes. The change in air pressure causes **inhalation** (breathing in) and **exhalation** (breathing out). These repeated breathing movements ventilate the lungs.

Gas	% by volume in inhaled air	% by volume in exhaled air
oxygen	21	16
carbon dioxide	0.035	4

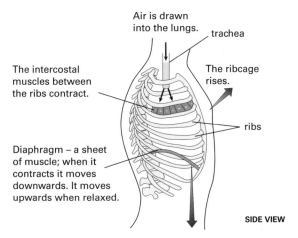

Inhalation: the diaphragm contracts and flattens.

Inhalation

- The diaphragm contracts and becomes less dome-shaped.
- At the same time the intercostal muscles between the ribs contract and raise the rib cage.
- The thoracic cavity enlarges. The resulting reduction in air pressure is transmitted via the **pleural cavity** to the lungs.
- The pressure of air in the alveoli is *less* than that of the atmosphere. Air, therefore, is drawn into the lungs through the trachea and bronchi.

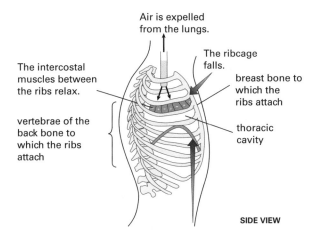

Exhalation: the diaphragm relaxes and curves upwards.

Exhalation

- The diaphragm and intercostal muscles relax, lowering the ribs and raising the diaphragm.
- The volume of the thoracic cavity decreases and the lungs are compressed.
- The pressure of air in the alveoli is *greater* than that of the atmosphere.
- The air passes from the lungs through the bronchi and trachea to the atmosphere.

The pleural cavity is the space between the lungs and the rib cage. It is lined by **pleural membranes**. The membranes are lubricated, facilitating (making easy) movements of the lungs. Gas pressure within the pleural cavity is less than that of the atmosphere.

31

Changes in the volume of inhaled and exhaled air

The volume of air exchanged between the lungs and the atmosphere depends on the body's activity and the extent and rate of ventilation (measured as the breathing rate) of the lungs. The volume of air breathed in and out is measured using a **spirometer**.

- The air capacity of the human lungs is about 5.5 dm^3, of which 1.5 dm^3 is the **residual volume**. The term refers to the air *not removed* from the lungs even when breathing is forced.
- At rest, normal breathing results in a **tidal volume** of around 0.5 dm^3. So a person with a resting breathing rate of 16 inhalations/exhalations per minute exchanges air at a rate of $16 \times 0.5 = 8$ dm^3 min^{-1}.
- Stress and strenuous physical activity increase the breathing rate and so increase the rate at which air is exchanged. The whole volume of the lungs less its residual volume comes into play. The volume is about 4.0 dm^3 (5.5 dm^3 − 1.5 dm^3) and represents the **vital capacity** of the lungs. For example, a breathing rate of 40 inhalations/exhalations per minute exchanges air at a rate of $40 \times 4.0 = 160$ dm^3 min^{-1}.

Changes in breathing rate help to keep the levels of oxygen and carbon dioxide in the blood constant.

Questions

3 Summarize the changes in air pressure in the lungs during one cycle of inhalation and exhalation.

4 Explain the relationship between residual volume, tidal volume, and vital capacity of the lungs.

The spirometer and the measurement of respiratory activity

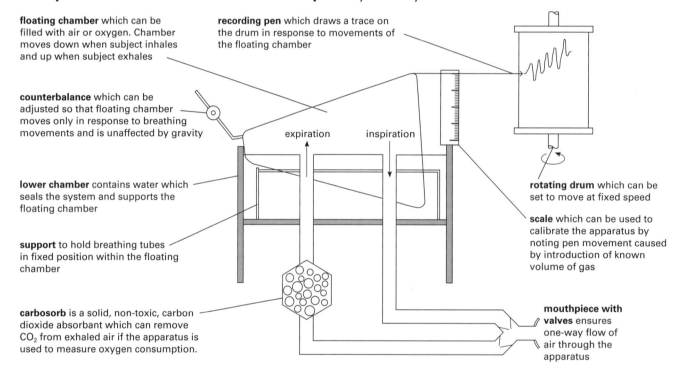

floating chamber which can be filled with air or oxygen. Chamber moves down when subject inhales and up when subject exhales

recording pen which draws a trace on the drum in response to movements of the floating chamber

counterbalance which can be adjusted so that floating chamber moves only in response to breathing movements and is unaffected by gravity

expiration   inspiration

lower chamber contains water which seals the system and supports the floating chamber

support to hold breathing tubes in fixed position within the floating chamber

carbosorb is a solid, non-toxic, carbon dioxide absorbant which can remove CO$_2$ from exhaled air if the apparatus is used to measure oxygen consumption.

rotating drum which can be set to move at fixed speed

scale which can be used to calibrate the apparatus by noting pen movement caused by introduction of known volume of gas

mouthpiece with valves ensures one-way flow of air through the apparatus

Measurement of lung volumes by spirometry

Vital capacity = maximum volume of air which can be exchanged from full inspiration to full expiration

Tidal volume = volume of air exchanged during normal quiet breathing

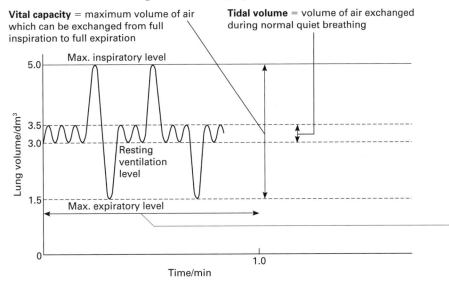

Pulmonary ventilation
(the volume of air available to the lungs per unit of time)

= tidal volume × breathing rate

P.V. can be increased by increasing either or both tidal volume and breathing rate

Breathing rate = number of breaths in one minute. About 14–16 bpm in a resting adult human

Measurement of oxygen uptake

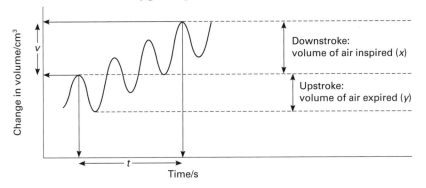

Downstroke: volume of air inspired (x)

Upstroke: volume of air expired (y)

Oxygen consumed
$$v = (x - y)$$

Rate of oxygen consumption
$$= \frac{v}{t} \text{ cm}^3 \text{ s}^{-1}$$

Arteries are often described as carriers of blood enriched with oxygen (oxygenated blood) and veins as carriers of blood depleted of oxygen (deoxygenated blood).

The need for a circulatory system

In a large organism the surface area: volume ratio becomes reduced, and so simple diffusion at its surface cannot supply the materials it needs. Such organisms have a system for transporting materials around their bodies.

- Oxygen is needed by all cells for respiration.
- Heat generated by metabolism must be distributed and transferred to the environment.
- Wastes such as urea and carbon dioxide must be removed from cells.
- Cells need nutrients such as glucose, amino acids, and mineral ions.

The larger and more active is an organism, the greater its transport needs.

Large organisms are multicellular – many small cells have a greater combined surface area than one large cell of the same total volume. As well as organ systems for gaseous exchange, they also have systems for transporting substances around their bodies – a **mass flow system**. In animals this is the **circulatory system**.

The human circulatory system – a double circulatory system

The heart pumps blood through tubular blood vessels (**arteries**, **veins**, and **capillaries**). Blood transports oxygen, digested food, hormones, and other substances to the tissues and organs of the body. It also carries carbon dioxide and other waste substances (e.g. urea) from the tissues and organs of the body to where they are removed from the body.

Remember that:

- arteries carry blood from the heart
- veins carry blood to the heart
- capillaries link arteries and veins

The heart and blood vessels are the components of the circulatory system.

Notice in the diagram on the next page that

- the **pulmonary arteries** carry deoxygenated blood (to the lungs from the heart) and the **pulmonary veins** carry oxygenated blood (from the lungs to the heart)
- unlike other veins, the **hepatic portal vein** does not drain blood into the **vena cava** *en route* to the heart, but carries blood with its load of digested food from the intestine to the liver

Double and single circulatory systems

The mammalian circulatory system is a **double circulatory system**. Blood flows through the heart twice for each complete circuit of the body. The pulmonary system (from the heart to the lungs and back) is separated from the systemic system (from the heart to the rest of the body and back). The two separate circuits allow rapid high-pressure distribution of oxygen in active, endothermic animals.

In contrast fish have a **single circulatory system**. Blood flows through the heart once for each complete circuit of the body.

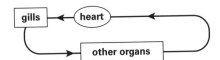

Blood leaving the gills is at a lower pressure and speed as it has flowed through the exchange organs with their many branching capillaries.

Open and closed circulatory systems

The circulatory systems of mammals and fish are **closed systems** – the blood is contained in blood vessels. Insects have a mass transport system that is an **open system**. Nutrients and wastes are transported in a fluid called hemolymph, which flows freely through the body and is in direct contact with organs and tissues. Muscle contractions push the hemolymph through the body.

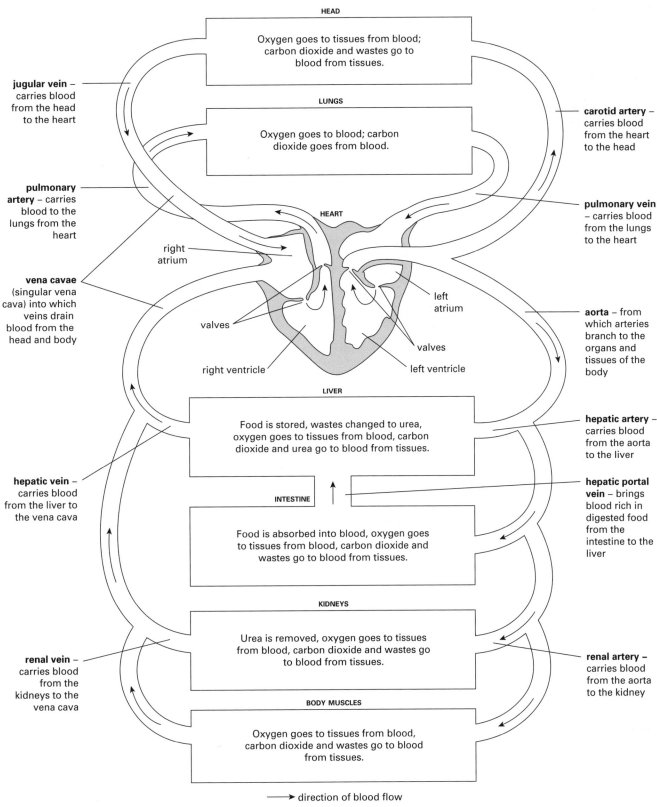

HEAD

Oxygen goes to tissues from blood; carbon dioxide and wastes go to blood from tissues.

LUNGS

Oxygen goes to blood; carbon dioxide goes from blood.

HEART

right atrium

left atrium

valves

valves

right ventricle

left ventricle

LIVER

Food is stored, wastes changed to urea, oxygen goes to tissues from blood, carbon dioxide and urea go to blood from tissues.

INTESTINE

Food is absorbed into blood, oxygen goes to tissues from blood, carbon dioxide and wastes go to blood from tissues.

KIDNEYS

Urea is removed, oxygen goes to tissues from blood, carbon dioxide and wastes go to blood from tissues.

BODY MUSCLES

Oxygen goes to tissues from blood, carbon dioxide and wastes go to blood from tissues.

jugular vein – carries blood from the head to the heart

pulmonary artery – carries blood to the lungs from the heart

vena cavae (singular vena cava) into which veins drain blood from the head and body

hepatic vein – carries blood from the liver to the vena cava

renal vein – carries blood from the kidneys to the vena cava

carotid artery – carries blood from the heart to the head

pulmonary vein – carries blood from the lungs to the heart

aorta – from which arteries branch to the organs and tissues of the body

hepatic artery – carries blood from the aorta to the liver

hepatic portal vein – brings blood rich in digested food from the intestine to the liver

renal artery – carries blood from the aorta to the kidney

⟶ direction of blood flow

The human double circulatory system

Heart structure and function

Heartbeat

Heartbeats are the result of the contraction and relaxation of the **cardiac** muscle of the heart. This action pumps blood through the system of blood vessels which make up the circulatory system.

Each beat is a two-tone sound made by the opening and closing of the valves which direct the flow of blood through the heart. A natural pacemaker located in the wall of the heart controls the heartbeat.

Heart structure

The heart lies in the chest cavity, protected by the rib cage. Inside it has four chambers, separated by a septum into the right-hand side and left-hand side. Outside, vessels supply blood to its surface. The diagrams show the arrangement.

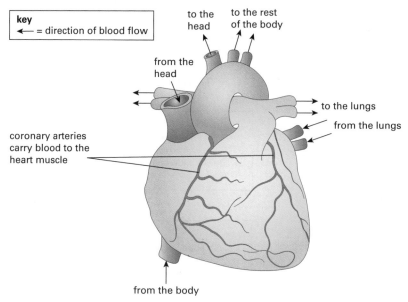

Blood supply to and from the heart (viewed from the front)

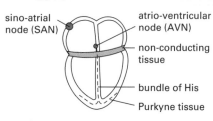

Coordinating the contraction of heart muscle. The SAN determines the resting rate of the heartbeat and is called the **pacemaker**.
Nerve impulses from the SAN to the AVN are prevented from reaching the ventricles by the band of non-impulse-conducting tissue in the wall of the heart.

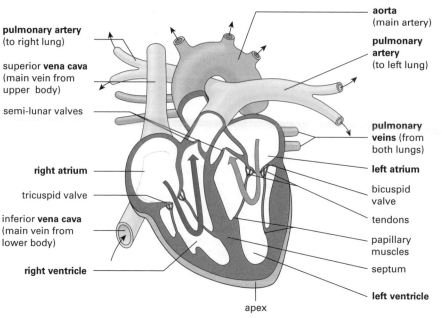

A vertical section through the heart (viewed from the front)

The cardiac cycle

The term **cardiac cycle** refers to the sequence of events which propel blood through the heart and its associated blood vessels. Use the diagram to follow the sequence.

Step 1 – Atrial diastole

- The atria are relaxed. The valves separating the atria from the ventricles are closed.
- The right atrium fills with deoxygenated blood from the venae cavae. The left atrium fills with oxygenated blood from the pulmonary veins.
- As the atria fill, increasing pressure is put on the valves. They start to open.

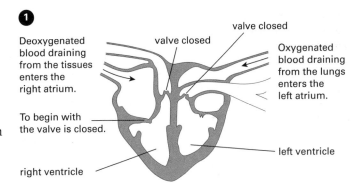

Step 2 – Atrial systole

- Nerve impulses generated in the **sino-atrial node (SAN)** spread out through the muscles of the atria.
- The atria contract. Their volume decreases so the pressure of blood inside increases.
- Blood is forced through the valves into the ventricles.
- The impulses from the SAN stimulate the atrio-ventricular node (AVN) in the septum separating the atria.
- Impulses from the AVN pass along the **Bundle of His** which is made up of strands of conductive tissue called **Purkyne tissue**.

Step 3 – Ventricular systole

- The wave of impulses from the bundle of His stimulates the muscle of the walls of the ventricles at their apex.
- The ventricles contract from the apex upwards. Their volume decreases so the pressure of the blood inside increases.
- This forces shut the valves separating the atria from the ventricles (preventing backflow into the atria), and opens the valves guarding the openings of the arteries.
- Blood is forced through the pulmonary artery and the aorta.
- The elasticity of the artery walls allows for the increase in the volume of blood. Back flow into the ventricles is prevented by the semi-lunar valves.

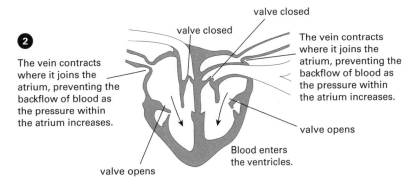

Step 4 – Ventricular diastole

- Relaxation of the ventricles marks the end of the cardiac cycle.

Step	Time in seconds
atrial systole	0.1
ventricular systole	0.3
atrial and ventricular diastole	0.4
Total time	**0.8**

Fact file

The heart can contract and relax rhythmically for a considerable time without the stimulus of nerve impulses or hormones. We say that the heart beat is **myogenic**. The term means that contraction originates in the heart itself.

Timings and pressure changes

The table shows the average timings during one cardiac cycle of a resting person. These timings and the corresponding changes in pressure in the atria, ventricles, and aorta are summarized in the diagram.

key
— = pressure in aorta
- - - = pressure in ventricle
······ = pressure in atrium

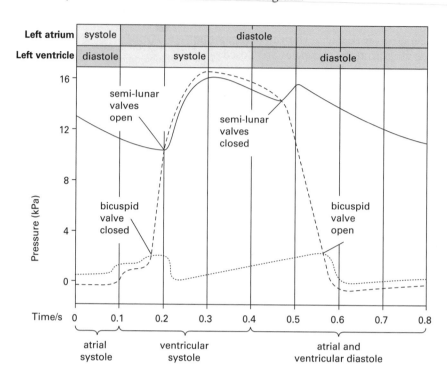

Pressure changes during the cardiac cycle. The maximal pressure in the left ventricle is greater than in the right ventricle because the wall of the left ventricle is thicker so it contracts more powerfully. The opening and closing of the different valves depends on the relative pressure on either side of each one.

Measuring heartbeat

The electrical activity of the heart's nerves and muscles can be detected by electrodes placed on the body's surface. The output is a series of waves in the form of a trace called an **ECG (electrocardiogram)**.

- The **P wave** records current flow through the atria from the SAN → AVN.
- The **QRS waves** record the spread of electrical activity through the ventricles.
- The **T wave** records the current generated following contraction of the ventricles.

An ECG is a picture of the heart's electrical activity and helps doctors diagnose its health.

ECG traces are interpreted by specialists. Some heart irregularities that an abnormal ECG trace may show include:

- missing P wave and irregular QRS waves – atrial fibrillation
- sawtooth P waves – atrial flutter.

The P wave indicates atrial depolarisation – the spread of an impulse from the SA node through the two atria. It is less powerful than the QRS complex since the atria are less massive than the ventricles.

The QRS wave (complex) represents ventricular depolarisation and is strong enough to mask atrial repolarisation.

The T wave represents ventricular repolarisation.

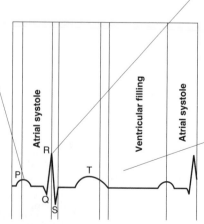

An ECG of a heartbeat

Questions

1 Describe the events of the cardiac cycle.
2 Name the waves which form the trace of an ECG.

Blood vessels

The diagram compares the structure and functions of arteries and veins.

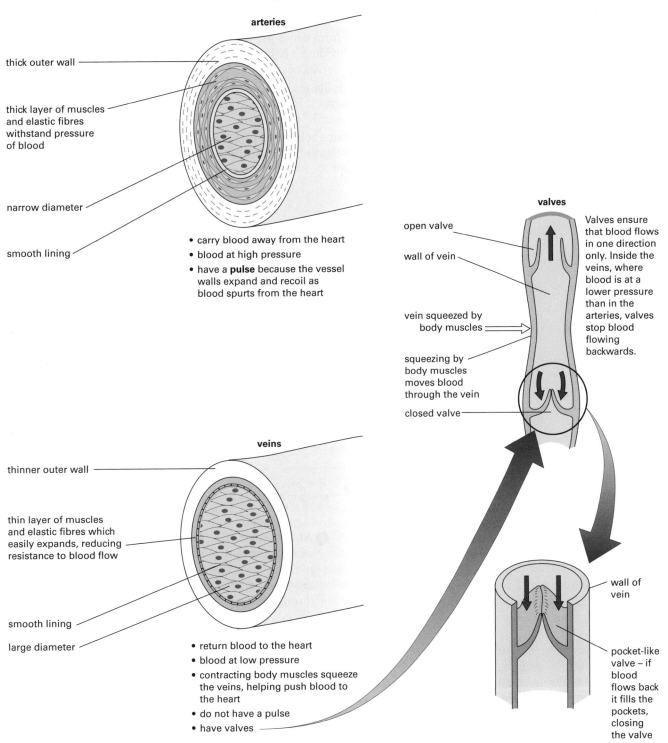

arteries

thick outer wall

thick layer of muscles and elastic fibres withstand pressure of blood

narrow diameter

smooth lining

- carry blood away from the heart
- blood at high pressure
- have a **pulse** because the vessel walls expand and recoil as blood spurts from the heart

valves

open valve

wall of vein

vein squeezed by body muscles

squeezing by body muscles moves blood through the vein

closed valve

Valves ensure that blood flows in one direction only. Inside the veins, where blood is at a lower pressure than in the arteries, valves stop blood flowing backwards.

veins

thinner outer wall

thin layer of muscles and elastic fibres which easily expands, reducing resistance to blood flow

smooth lining

large diameter

- return blood to the heart
- blood at low pressure
- contracting body muscles squeeze the veins, helping push blood to the heart
- do not have a pulse
- have valves

wall of vein

pocket-like valve – if blood flows back it fills the pockets, closing the valve

Notice in the diagram:
- Blood in veins is at a lower pressure than blood in arteries.
- One-way valves inside the veins prevent blood from flowing backwards.
- The propulsive force of the heart beat keeps blood flowing away from the heart through the arteries, so there is no need for valves inside arteries.

Arteries and veins branch into smaller vessels.
- Arteries branch into **arterioles**.
- Veins branch into **venules**.
- Arterioles and venules branch further into microscopic **capillaries**.

Exchanges between capillaries and tissues, and the role of lymph

Remember:

- The walls of capillary blood vessels are one cell thick.
  - ® As a result substances easily diffuse between blood in the capillaries and the surrounding tissues
- The capillaries form dense networks called **capillary beds** in the tissues of the body, providing a *large surface area* which maximizes the rate of exchange of materials between the blood and tissues.
- The blood in capillaries supplies nearby cells with oxygen, food molecules, and other substances. It also carries away carbon dioxide and other waste produced by the cells' metabolism.

Refer to the diagrams below and on the next page as you read about the exchanges of substances between capillaries and tissues, and the movement of fluid into the lymphatic system.

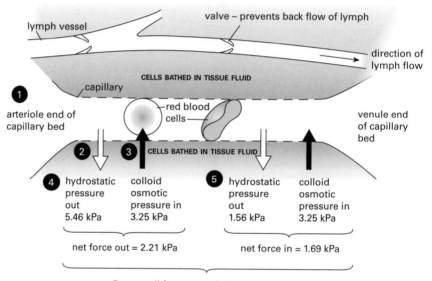

Exchanges between capillaries and tissues, and their relationship with the lymphatic system

❶ Hydrostatic pressure at the arteriole end of a capillary bed is high because of the force generated by contractions of the heart.

❷ The pressure forces small molecules dissolved in the blood plasma through the walls of the capillaries into surrounding tissues. The plasma is now called **tissue fluid**.

❸ Water escapes through the walls of the capillaries but not large protein molecules. So the water potential of the blood is lowered (more negative). This water potential is called the **colloid osmotic pressure** and has the effect of drawing molecules back into the blood capillaries.

❹ At the arteriole end of a capillary bed, the hydrostatic pressure forcing molecules out of the capillary vessels is greater than that of the colloidal osmotic pressure drawing them in.

- ® As a result there is a net outflow of substances in solution from the capillaries.

❺ At the venule end of the capillary bed, the colloid osmotic pressure is greater than the hydrostatic pressure – which is now reduced because of the resistance of the capillary walls to the flow of blood through the capillary vessels.

- ® As a result there is a net movement of substances in solution into the capillaries.

❻ The movement of substances in solution out of the capillaries is greater than the return flow.

- ® As a result an excess of fluid bathes the tissues.

❼ The excess tissue fluid drains into the **lymph vessels** which pass to all of the tissues of the body, as the diagram shows. The tissue fluid is now called **lymph**.

❽ The system of lymph vessels joins the blood system at the opening of the thoracic duct. Lymph continually circulates to the blood at this point.

- ® As a result the volume of the lymph in the lymph vessels remains constant.

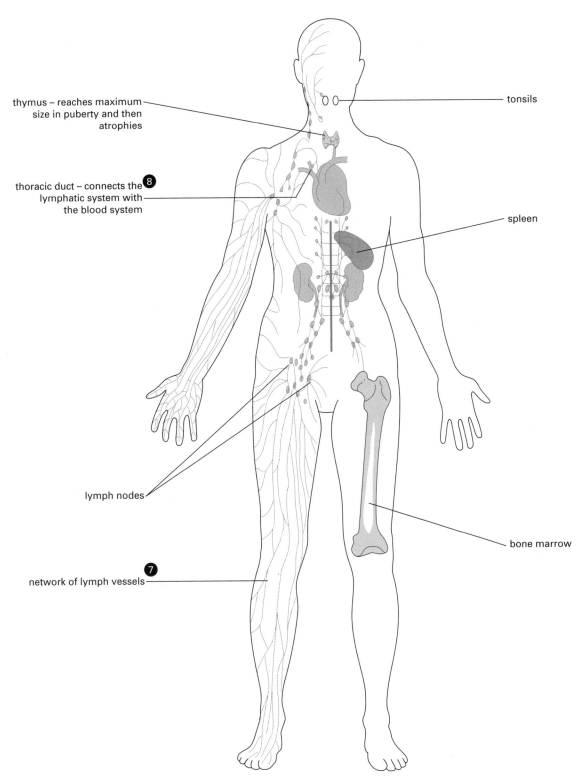

thymus – reaches maximum size in puberty and then atrophies

tonsils

thoracic duct – connects the lymphatic system with the blood system **8**

spleen

lymph nodes

bone marrow

network of lymph vessels **7**

The network of lymph vessels is shown on the right-hand side of the body only. The tonsils, thymus, and spleen are part of the lymphatic system.

Questions

1 Identify some of the differences between arteries, veins, and capillary blood vessels.

2 Briefly describe the role of valves in veins.

3 Summarize the processes which lead to the exchange of substances between capillary blood vessels and tissues.

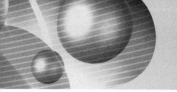

The haemoglobin molecule

Haemoglobin is the oxygen-carrying pigment in red blood cells. It is a **globular** protein with a quaternary structure, consisting of

- four coiled polypeptide chains (the **globin** part of the molecule)
- four **haem** groups

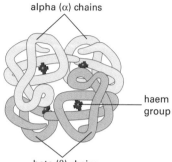

alpha (α) chains

haem group

beta (β) chains

The haemoglobin molecule

> Most enzymes, antibodies, and some hormones are globular proteins made of polypeptide chains that fold into a spherical shape. They are usually soluble in water.

Each haem group is a **prosthetic group** – part of the protein but not made of amino acids.

- The four haem groups each contain an iron ion that can combine with a molecule of oxygen.
  - As a result a molecule of haemoglobin can combine with four molecules of oxygen.

A haem group of haemoglobin

Fact file

As well as transporting oxygen to the respiring cells, haemoglobin also carries about 10% of the waste carbon dioxide that is transported by the blood to the alveoli. Carbon dioxide combines with the globin part of the molecule to form carbaminohaemoglobin.

Haemoglobin transports oxygen

Oxygen combines with haemoglobin forming **oxyhaemoglobin** in tissues where the concentration of oxygen is high. It quickly releases its oxygen in tissues where the concentration of oxygen is low.

Haemoglobin is therefore ideal for the transport of oxygen from the lungs (where the concentration of oxygen is high) to the rest of the body's tissues (where the concentration of oxygen may be low).

$$\text{haemoglobin} + \text{oxygen} \underset{\text{other body tissues}}{\overset{\text{lungs}}{\rightleftarrows}} \text{oxyhaemoglobin}$$

- Blood which contains a lot of oxyhaemoglobin is called **oxygenated** blood and is bright red.
- Blood with less oxyhaemoglobin in it is called **deoxygenated** blood and looks red/purple.

Uptake of oxygen and the dissociation curve

The partial pressure (in kPa) of a gas is a measure of its concentration. It is proportional to its percentage by volume in a mixture of gases.

- The atmosphere contains nearly 21% oxygen. The partial pressure of oxygen is therefore about 21 kPa.
- The combination of haemoglobin with oxygen depends on the partial pressure of oxygen in contact with it.

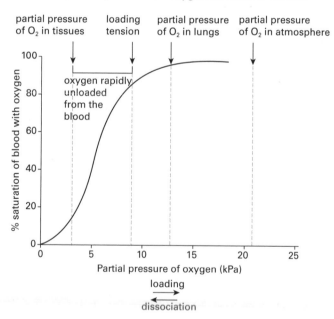

Oxygen association/dissociation curve for adult haemoglobin

Reading the graph from left to right shows the relationship between the *uptake* of oxygen by haemoglobin (association) and the increasing partial pressure (concentration) of oxygen in contact with it. Notice:

- Haemoglobin takes up oxygen rapidly for a relatively small increase in the partial pressure of the gas. The term **loading tension** refers to the point when 95% of the pigment is saturated.

- The loading tension corresponds to a partial pressure which is considerably less than the partial pressure of oxygen in the atmosphere.
  - As a result the blood supplying the lungs becomes rapidly loaded with oxygen at the partial pressure of oxygen normally found in the lungs.

Reading the graph from right to left shows the relationship between the *release* of oxygen by haemoglobin (dissociation) and the decreasing partial pressure of oxygen in contact with it.

Notice:

- Haemoglobin releases (unloads) oxygen rapidly for a relatively small decrease in partial pressure of the gas.
- The unloading of oxygen corresponds to a partial pressure of oxygen normally found in tissues which are using oxygen in aerobic respiration.
  - As a result tissues receive enough oxygen for their activities.

What makes the dissociation curve S-shaped?

Remember that a haemoglobin molecule combines with four molecules of oxygen.

- The combination of oxygen with one haem group slightly changes the shape of the haemoglobin molecule.
- The shape change makes it easier for one of the other haem groups to load an oxygen molecule.
- The combination of oxygen with this haem group makes it even easier for a third haem group to load oxygen, and so on.

The S-shape of the dissociation curve is the result of these knock-on effects.

The effects work in reverse:

- As the partial pressure of oxygen decreases, an oxygen molecule may be released.
- Its loss slightly changes the shape of the haemoglobin molecule, making the unloading of subsequent oxygen molecules increasingly easy.

How do the loading and unloading properties of haemoglobin help with oxygen transport?

The partial pressure of oxygen in the lungs is about 12 kPa. In the tissues the partial pressure of oxygen is much lower – 2 kPa in muscle tissue for example. Haemoglobin exposed to this concentration of oxygen is less than 20% saturated. The oxygen released from the haemoglobin molecules passes into the blood plasma, diffuses into the muscle cells and is used in aerobic respiration.

The Bohr effect

- Reducing the partial pressure of carbon dioxide *increases* the oxygen load of the blood for a given partial pressure of oxygen.
- Increasing the partial pressure of carbon dioxide *reduces* the oxygen load of the blood for a given partial pressure of oxygen.
- Increasing temperature has a similar effect.

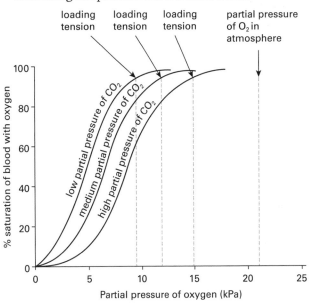

The more its dissociation curve is shifted to the right, the *less* readily haemoglobin loads oxygen, and the *more* readily it releases it. The more the curve is shifted to the left, the *more* readily haemoglobin loads oxygen and the *less* readily it releases it.

These alterations in the oxygen dissociation curve (called the **Bohr effect**) help to adjust the amount of oxygen tissues receive. For example, the demand for oxygen by muscle tissues during vigorous exercise is high. As the tissues respire more and more carbon dioxide is released. The temperature of the tissues also increases. These changes in the micro-environment of the tissues cause the dissociation curve to shift to the right, increasing the supply of oxygen to the tissues at a time when it is needed.

Fetal haemoglobin

The blood of the fetus is separated from that of its mother. Fetal and maternal blood vessels run alongside each other in the placenta where oxygen, carbon dioxide, nutrients, and urea are exchanged between them.

Fetal haemoglobin has a higher affinity for oxygen than does adult haemoglobin. This means that the dissociation curve for fetal haemoglobin lies to the left of the curve for adult haemoglobin, ensuring that oxygen passes from the mother's blood to that of the fetus in the placenta.

Questions

1 Why is haemoglobin described as a globular protein with a quaternary structure?

In any large organism the surface area: volume ratio is reduced, and simple diffusion at its surface cannot supply the materials it needs. Multicellular plants need to transport

- oxygen to all cells for respiration
- carbon dioxide to photosynthesising cells
- water and mineral ions from the soil, through the root throughout the plant
- glucose and other products of photosynthesis throughout the plant
- wastes to be removed from cells and from the plant

Just as with an animal, the larger the plant, the greater its transport needs. Multicellular plants have a **mass flow system** which comprises the xylem and phloem.

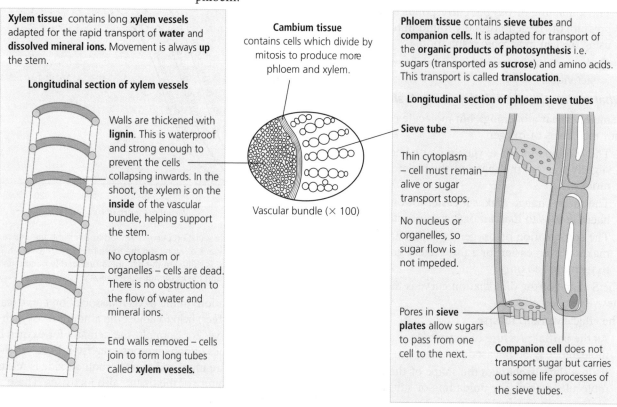

Xylem tissue contains long **xylem vessels** adapted for the rapid transport of **water** and **dissolved mineral ions**. Movement is always **up** the stem.

Longitudinal section of xylem vessels

Walls are thickened with **lignin**. This is waterproof and strong enough to prevent the cells collapsing inwards. In the shoot, the xylem is on the **inside** of the vascular bundle, helping support the stem.

No cytoplasm or organelles – cells are dead. There is no obstruction to the flow of water and mineral ions.

End walls removed – cells join to form long tubes called **xylem vessels**.

Cambium tissue contains cells which divide by mitosis to produce more phloem and xylem.

Vascular bundle (× 100)

Phloem tissue contains **sieve tubes** and **companion cells**. It is adapted for transport of the **organic products of photosynthesis** i.e. sugars (transported as **sucrose**) and amino acids. This transport is called **translocation**.

Longitudinal section of phloem sieve tubes

Sieve tube

Thin cytoplasm – cell must remain alive or sugar transport stops.

No nucleus or organelles, so sugar flow is not impeded.

Pores in **sieve plates** allow sugars to pass from one cell to the next.

Companion cell does not transport sugar but carries out some life processes of the sieve tubes.

Direction of transport varies with the seasons!
Sucrose is transported **from** stores in the root **to** leaves in spring, but **to** stores in the root **from** photosynthesising leaves in the summer and early autumn. Whatever the time of year the movement of sugars and amino acids (translocation) is from **source** to **sink**. In other words, sucrose and amino acids are translocated from the region where they are made or absorbed to the region where they are stored or used.

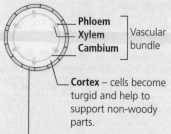

Stem – vascular bundles are arranged in a ring with soft cortex in the centre, helping to support the stem.

Phloem
Xylem   Vascular
Cambium   bundle

Cortex – cells become turgid and help to support non-woody parts.

Epidermis – protects against infection by viruses and bacteria, and dehydration.

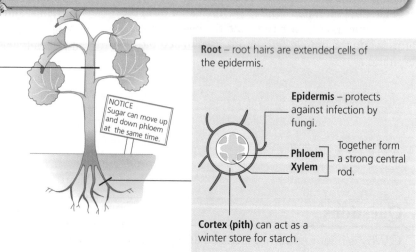

NOTICE
Sugar can move up and down phloem at the same time.

Root – root hairs are extended cells of the epidermis.

Epidermis – protects against infection by fungi.

Phloem   Together form
Xylem   a strong central
               rod.

Cortex (pith) can act as a winter store for starch.

The uptake and passage of water across the root

Root hairs are in intimate contact with soil particles. Water passes into the root by osmosis.

Some of the water takes the **apoplastic route**:

- Water passes into spaces between cellulose fibres within the cell walls of root hair cells.
- The water passes from root hair cells across the root from cell wall to cell wall.
- Its passage is due to the pull transmitted by the cohesive forces between water molecules as a result of hydrogen bonding.

Water also takes the **symplastic route** and **vacuolar route** across a root:

- Water passes from where its potential is high in the soil to where its potential is lower in the cells of the cortex.
- The difference in water potential (water potential gradient) between adjacent cells means that water moves by osmosis through the root hair cells and through the cells of the cortex to the xylem.

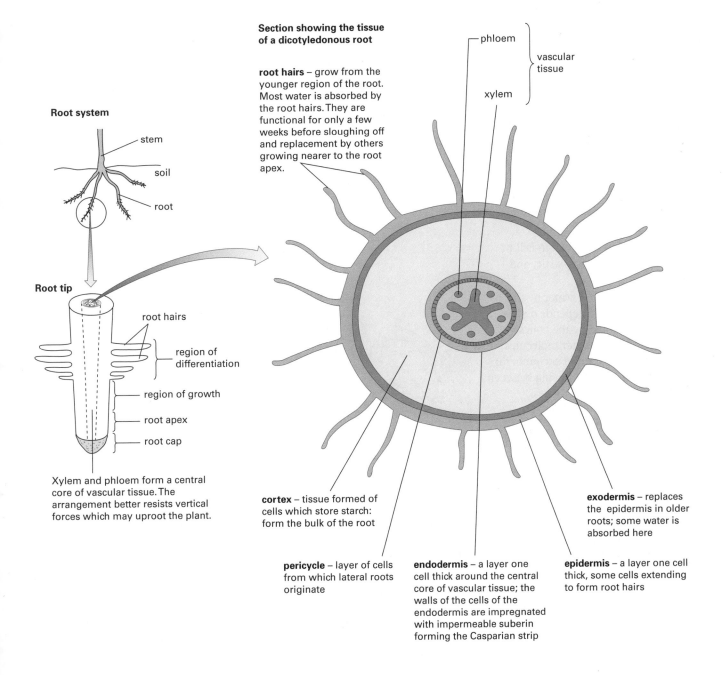

Section showing the tissue of a dicotyledonous root

phloem ⎫
         ⎬ vascular tissue
xylem   ⎭

root hairs – grow from the younger region of the root. Most water is absorbed by the root hairs. They are functional for only a few weeks before sloughing off and replacement by others growing nearer to the root apex.

Root system

stem

soil

root

Root tip

root hairs

region of differentiation

region of growth

root apex

root cap

Xylem and phloem form a central core of vascular tissue. The arrangement better resists vertical forces which may uproot the plant.

cortex – tissue formed of cells which store starch: form the bulk of the root

exodermis – replaces the epidermis in older roots; some water is absorbed here

pericycle – layer of cells from which lateral roots originate

endodermis – a layer one cell thick around the central core of vascular tissue; the walls of the cells of the endodermis are impregnated with impermeable suberin forming the Casparian strip

epidermis – a layer one cell thick, some cells extending to form root hairs

45

- In the symplastic route water diffuses through the cytoplasm of *adjacent* cells.
- In the vacuolar route water diffuses through the *vacuoles* as well as the cytoplasm.

The impermeable **Casparian strip** seems to prevent the passage of water by the apoplastic route. Water therefore passes into the cytoplasm and vacuoles of the cells of the endodermis on its way to the xylem.

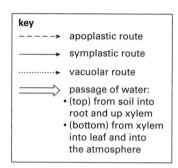

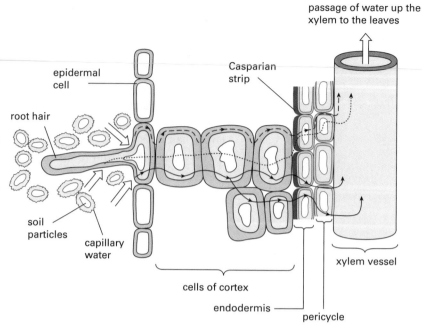

The passage of water across the leaf

The diagram shows the passage of water across the tissues of the leaf and the loss of water from the air spaces of the leaf to the atmosphere. Notice:

- Water moves from cell to cell by the apoplastic, symplastic, and vacuolar routes.
- The movement of water by the apoplastic route depends on the pull transmitted by the cohesive forces between water molecules. The apoplastic route accounts for most of the water moving between the cells of the leaf tissues.
- The movement of water by the symplastic and vacuolar routes depends on the difference in water potential (water potential gradient) between adjacent cells.
- The loss of water from the leaf depends on the difference in water potential (water potential gradient) between its air spaces and the atmosphere outside.

Overall there is a water potential gradient across the leaf from the leaf xylem to the atmosphere. The loss of water from the leaf through the stomata into the atmosphere is called **transpiration**.

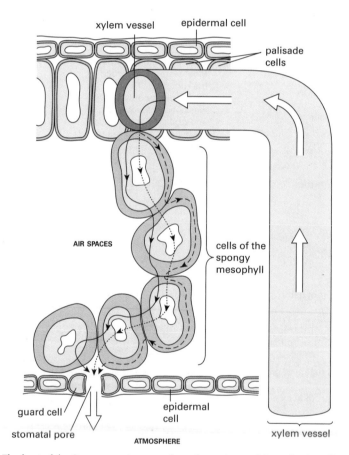

The heat of the Sun evaporates water from the surfaces of the palisade cells and spongy mesophyll cells. The water vapour saturates the air spaces of the leaf.

The movement of water up the xylem of the stem

As water is lost from the leaf through transpiration it is replaced by more water drawn by osmosis from the xylem of the leaf into the adjacent mesophyll cells.

The movement of water molecules into the tissues of the leaf pulls (draws up) other water molecules through the xylem of the stem. This is because of the pull of water molecules moving from wall to wall of the cells of the leaf's tissues by the apoplastic route. The effect is called **transpiration pull** and produces a state of tension in the columns of water within the xylem vessels.

Transpiration pull is possible because of the considerable cohesive forces between water molecules as a result of hydrogen bonding. These cohesive forces are sufficient to raise water to the tops of the tallest trees, and the theory of the mechanism is known as the **cohesion-tension theory**.

In addition, **adhesion** is an attractive force between the water molecules and the walls of the xylem tubes. Adhesion also contributes to maintaining the transpiration stream.

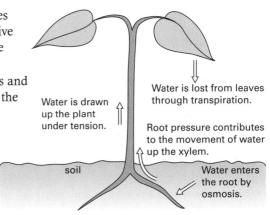

Water is lost from leaves through transpiration.

Water is drawn up the plant under tension.

Root pressure contributes to the movement of water up the xylem.

soil

Water enters the root by osmosis.

Movement of water in a whole plant

In summary

Water movement through the plant occurs because of

- the difference in water potential between the soil water and root tissue, and between leaf tissue and the atmosphere
- transpiration pull which produces tension in the columns of water within the xylem of the stem (made possible because of the cohesion of water molecules through hydrogen bonding)

Transpiration pull also reduces the hydrostatic pressure at the base of the xylem compared with the hydrostatic pressure which develops as water is drawn across the root. This difference in hydrostatic pressure is responsible for the entry of water into the xylem.

The forces generated as water is drawn across the root cause a build up of **root pressure**. This contributes to the movement of the water up the xylem of the stem, especially in the relatively short **herbaceous** (non-woody plants). However it does not account for the movement of water up tall trees.

Rate of transpiration

The table below shows how different factors affect the rate of transpiration. These factors cause changes in the

- concentration gradient of water vapour between the inside of the leaf and the atmosphere outside (cause **A**)
- size of the aperture of the stomatal pores (cause **B**)
- kinetic motion of water molecules (cause **C**)

Factor	Rate of transpiration		Cause
	Increase	**Decrease**	
humidity	low	high	A – The steeper the concentration gradient, the faster is the rate of transpiration.
wind condition	windy	still	
light	bright	dim	B – The larger the aperture, the faster is the rate of transpiration.
temperature	high	low	C – An increase in temperature increases the kinetic motion of water molecules. The greater their kinetic motion, the greater is the rate of transpiration.

Fact file

Transpiration is a consequence of respiration. The stomata in the leaves open in sunlight to allow carbon dioxide to enter the leaf for photosynthesis. Water evaporates through the open stomata, driving transpiration.

Diffusion shells

There is a layer of stationary air called a **diffusion shell** adjacent to the leaf's surface. The thickness of the diffusion shell depends on the structure of the leaf, e.g. its size and shape and whether it is hairy. It also depends on the wind speed. Water vapour diffuses across the diffusion shell before being carried away by moving air. Any factor that reduces the thickness of the diffusion shell, such as wind speed, increases the rate of evaporation of water vapour from the leaf.

Mass transport

Systems of **mass transport** are needed to move substances rapidly from one part of a large organism to another.

The systems link with exchange surfaces where differences in concentration of substances and in pressure of gases propel substances

- from where they are exchanged
- to where they are needed by tissues or removed to the environment.

The blood system of mammals and the xylem and phloem tissues of plants are examples of mass transport systems.

Questions

1 Explain the roles of the endodermis and pericycle in the root of a dicotyledonous plant.

2 The terms 'apoplastic route' and 'symplastic route' refer to the passage of water across the roots and leaves of a dicotyledonous plant. Explain the difference between the two terms.

3 Summarize the processes by which water passes from the soil, into a plant, through the plant and into the atmosphere.

The bubble potometer

The potometer measures water uptake (= water loss by transpiration + water consumption for cell expansion and photosynthesis).

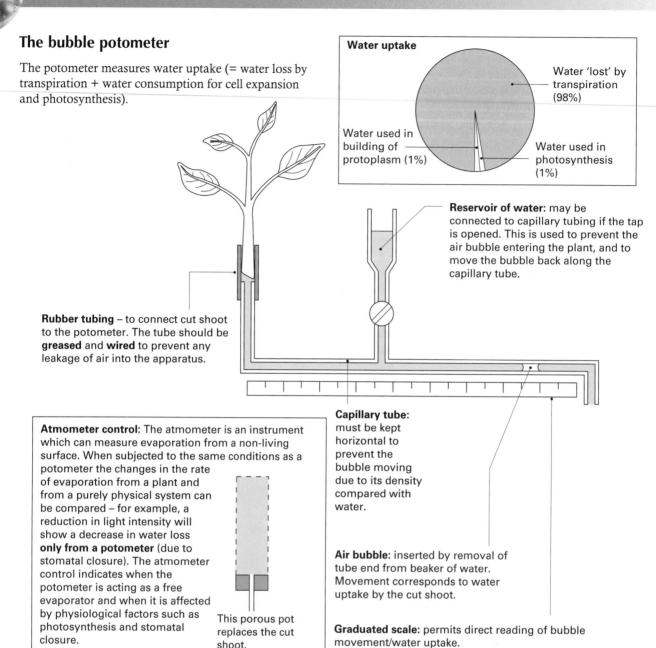

Water uptake

Water 'lost' by transpiration (98%)

Water used in building of protoplasm (1%)

Water used in photosynthesis (1%)

Reservoir of water: may be connected to capillary tubing if the tap is opened. This is used to prevent the air bubble entering the plant, and to move the bubble back along the capillary tube.

Rubber tubing – to connect cut shoot to the potometer. The tube should be **greased** and **wired** to prevent any leakage of air into the apparatus.

Atmometer control: The atmometer is an instrument which can measure evaporation from a non-living surface. When subjected to the same conditions as a potometer the changes in the rate of evaporation from a plant and from a purely physical system can be compared – for example, a reduction in light intensity will show a decrease in water loss **only from a potometer** (due to stomatal closure). The atmometer control indicates when the potometer is acting as a free evaporator and when it is affected by physiological factors such as photosynthesis and stomatal closure.

This porous pot replaces the cut shoot.

Capillary tube: must be kept horizontal to prevent the bubble moving due to its density compared with water.

Air bubble: inserted by removal of tube end from beaker of water. Movement corresponds to water uptake by the cut shoot.

Graduated scale: permits direct reading of bubble movement/water uptake.

Procedure

1. The leafy shoot must be cut **under water**, the apparatus must be filled **under water** and the shoot fixed to the potometer **under water** to prevent air locks in the system.

2. Allow plant to equilibrate (5 min) before introduction of air bubble. Take at least three readings of rate of bubble movement, and use reservoir to return bubble to zero on each occasion. Calculate mean of readings. Record air temperature.

3. Scale can be calibrated by introducing a known mass of mercury into the capillary tubing and using $\rho = m/v$ (density ρ for mercury is known, m can be measured, thus v corresponding to a measured distance of bubble movement can be determined).

4. Rate of water uptake per unit area of leaves can be calculated by measurement of leaf area.

External factors affecting transpiration

- **Light intensity:** use bench lamp (with water bath to act as heat filter) to increase light intensity. To simulate 'darkness' enclose shoot in black polythene bag.

- **Humidity:** enclose shoot in clear plastic bag to **increase** relative humidity of atmosphere – include water absorbant such as calcium chloride to **decrease** relative humidity.

- **Wind:** use small electric fan with 'cool' control to mimic air movements whilst avoiding effects of temperature changes.

- May also determine relative importance of upper surface/lower surface/stem/petiole in water loss by smearing with vaseline (acts like a waxy cuticle) as appropriate.

N.B. It is sometimes difficult to change only one condition at a time, e.g. enclosure in a black bag to eliminate light will also increase the relative humidity of the atmosphere.

Gas exchange across the leaves of plants

The diagram shows how different adaptations maximize the exchange of gases between the leaves of a plant and the environment.

Fact file

The **cotyledon** is the part of a seed in which starch is stored. The starch is a source of energy during germination, when the embryo plant grows and develops.

Seeds with two cotyledons are called **dicotyledonous** seeds. The leaves of plants whose seeds are dicotyledonous are broad – the so-called broad-leaved plants.

The surfaces of the cells of the palisade and spongy mesophyll tissues are gas exchange surfaces in the leaf. They are moist and permeable. Gases in solution are quickly exchanged across the cell surfaces.

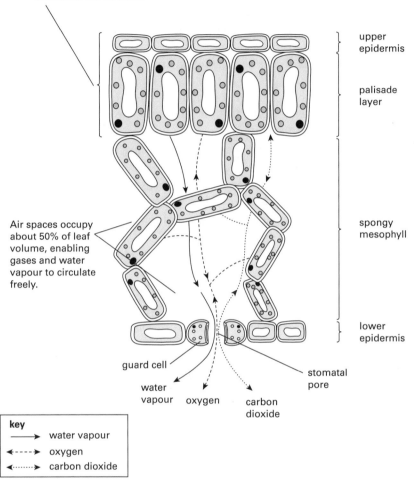

Air spaces occupy about 50% of leaf volume, enabling gases and water vapour to circulate freely.

upper epidermis

palisade layer

spongy mesophyll

lower epidermis

guard cell

water vapour   oxygen   carbon dioxide

stomatal pore

key
→ water vapour
◄----► oxygen
◄·······► carbon dioxide

The gas exchange surfaces inside the leaf of a dicotyledonous plant

Notice that the under surface of the leaf is perforated with gaps called stomata. Each **stoma** (singular) is flanked by **guard cells** which contain chloroplasts.

- Recall that the direction of the net movement of oxygen and carbon dioxide molecules between the inside of a leaf and the atmosphere depends on the balance between the rate of photosynthesis and the rate of aerobic respiration of the leaf's tissues.
- When the concentration of carbon dioxide produced in aerobic respiration balances that used in photosynthesis, then the net exchange of carbon dioxide between the leaf and the atmosphere is zero. This is the **compensation point**.

Recall also that

- the spaces inside a leaf are saturated with water vapour
- the concentration of water vapour in the atmosphere is usually less than inside the leaf.

- As a result water vapour passes down its concentration gradient from inside the leaf, through the stomata perforating the under surface of the leaf to the atmosphere. The process is called **transpiration**.

Fact file

Any characteristic which increases an organism's chances of surviving in its habitat is called an **adaptation**. The term **xerophytic** refers to plants able to survive in hot, dry conditions.

Cacti are xerophytic plants. The adaptations of cacti are a compromise between the opposing requirements of gas exchange and the limitation of the loss of water from the plant to the environment.

Leaves are pointed **spines**, reducing their surface area and so reducing water loss. They also discourage animals from eating the cactus.

A thick **waxy waterproof layer** covering the surfaces reduces water loss.

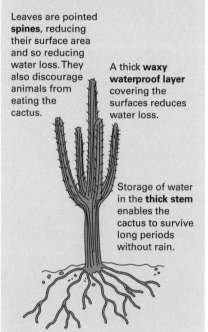

Storage of water in the **thick stem** enables the cactus to survive long periods without rain.

Long roots branch out just under the sand's surface. The **large surface area** of the roots enables the cactus to **maximize absorption** of what little water is available.

Other examples of xerophytic adaptations include stomata sunken into deep pits, and leaves rolled with the stomata inside. Xerophytic adaptations increase the size of diffusion shells around the leaf, so reducing the evaporation of water vapour from the leaf.

How stomata open and close

- When guard cells fill with water and become more turgid, their volume increases.
  - As a result the guard cells push each other apart and the pore opens.
- Guard cells become more turgid because of the active transport of potassium ions (K^+) into the cells from the surrounding cells of the leaf's lower epidermis. Active transport removes chloride ions (Cl^-) from the guard cells at the same time.
  - As a result of the exchange of ions, the water potential of the guard cells becomes more negative compared with the cells surrounding them.
  - As a result water passes into the guard cells from the surrounding cells by osmosis.
  - As a result the guard cells become turgid and bow outwards, opening the pore.
- The active transport of potassium ions and chloride ions is triggered by light and photosynthesis. Recall that the guard cells have chloroplasts; other cells of the leaf epidermis do not.
- In the dark (and absence of photosynthesis) the active transport of ions stops.
  - As a result the ions diffuse down their respective concentration gradients until equilibrium with the surrounding epidermal cells is reached.
  - As a result the water potential of the guard cells becomes less negative compared with the cells surrounding them.
  - As a result water passes from the guard cells into the surrounding cells by osmosis.
  - As a result the guard cells become less turgid and the pore closes.

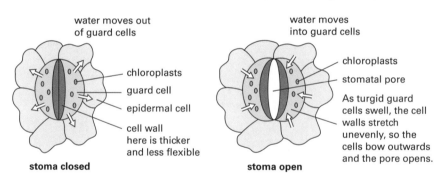

water moves out of guard cells

chloroplasts
guard cell
epidermal cell
cell wall here is thicker and less flexible

stoma closed

water moves into guard cells

chloroplasts
stomatal pore
As turgid guard cells swell, the cell walls stretch unevenly, so the cells bow outwards and the pore opens.

stoma open

How stomata open and close

Questions

1 Summarize the adaptations of gas exchange surfaces which maximize the rate of diffusion across them.

2 Explain why concentration gradients of oxygen and carbon dioxide across the surface of the single-celled organism *Amoeba proteus* enable it to exchange the gases between itself and its environment.

3 Summarize the process which controls the opening and closing of the stomata of a leaf.

1.20 Translocation

Fact file

Translocation is an energy-requiring process that transports substances made in the plant, e.g. sucrose, between sources and sinks.

Mass flow theory describes transport in the phloem: movements of water and sucrose generate a gradient of hydrostatic pressure which drives the translocation of organic solutes from leaves (the 'source' of solutes) to roots and meristems (the 'sink' for solutes). As well as being sinks for organic solutes made in the leaves, roots also act as storage organs. In this case they act as sources.

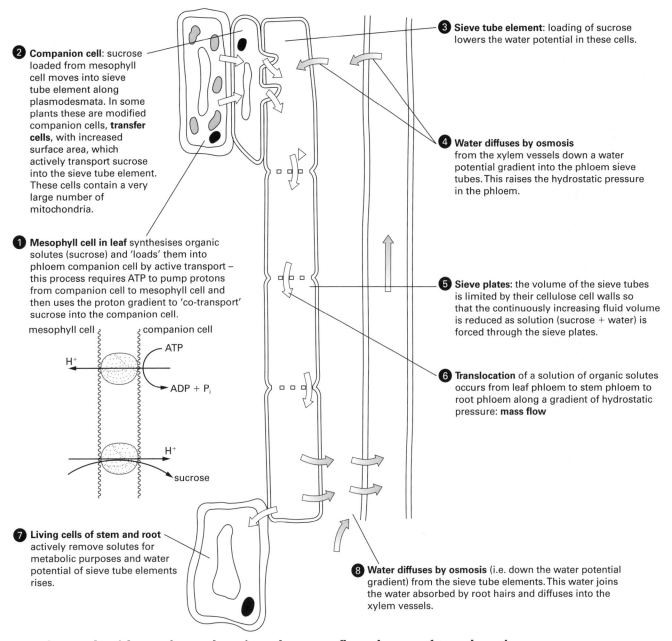

❷ **Companion cell:** sucrose loaded from mesophyll cell moves into sieve tube element along plasmodesmata. In some plants these are modified companion cells, **transfer cells**, with increased surface area, which actively transport sucrose into the sieve tube element. These cells contain a very large number of mitochondria.

❶ **Mesophyll cell in leaf** synthesises organic solutes (sucrose) and 'loads' them into phloem companion cell by active transport – this process requires ATP to pump protons from companion cell to mesophyll cell and then uses the proton gradient to 'co-transport' sucrose into the companion cell.

mesophyll cell   companion cell

H^+

ATP

$ADP + P_i$

H^+

sucrose

❼ **Living cells of stem and root** actively remove solutes for metabolic purposes and water potential of sieve tube elements rises.

❸ **Sieve tube element:** loading of sucrose lowers the water potential in these cells.

❹ **Water diffuses by osmosis** from the xylem vessels down a water potential gradient into the phloem sieve tubes. This raises the hydrostatic pressure in the phloem.

❺ **Sieve plates:** the volume of the sieve tubes is limited by their cellulose cell walls so that the continuously increasing fluid volume is reduced as solution (sucrose + water) is forced through the sieve plates.

❻ **Translocation** of a solution of organic solutes occurs from leaf phloem to stem phloem to root phloem along a gradient of hydrostatic pressure: **mass flow**

❽ **Water diffuses by osmosis** (i.e. down the water potential gradient) from the sieve tube elements. This water joins the water absorbed by root hairs and diffuses into the xylem vessels.

Experimental evidence for and against the mass flow theory of translocation

Evidence for	Evidence against
If a stem is punctured by an aphid and the aphid removed, leaving its mouthparts (stylet) in the stem, sap exudes through the stylet under pressure. This can be analysed and shown to contain sucrose and amino acids.	The plant can be exposed to radioactively labelled carbon dioxide. Biologists can then measure how long it takes for labelled sucrose and other substances to appear in the phloem, and how long it takes for them to be transported down the phloem. Not all solutes are carried at the same speed, and this difference cannot be explained by the mass flow theory.
Concentration gradients between sources and sinks have been measured, which supports the mass flow theory.	Solutes have been measured moving in both directions (up and down the phloem), which does not support the mass flow theory.

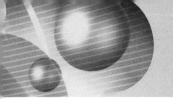

Physical properties of water

Water's unusual physical properties are explained by hydrogen bonding between the individual molecules.

Solvent properties: the polarity of water makes it an excellent solvent for other polar molecules ...

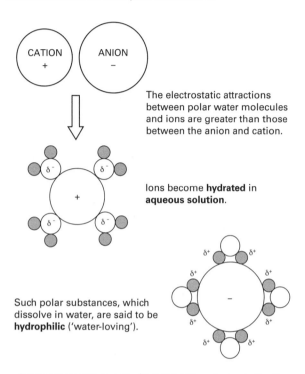

The electrostatic attractions between polar water molecules and ions are greater than those between the anion and cation.

Ions become **hydrated** in **aqueous solution**.

Such polar substances, which dissolve in water, are said to be **hydrophilic** ('water-loving').

... but means that non-polar (**hydrophobic** or 'water-hating') substances do not readily dissolve in water.

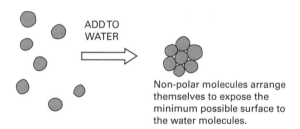

Non-polar molecules arrange themselves to expose the minimum possible surface to the water molecules.

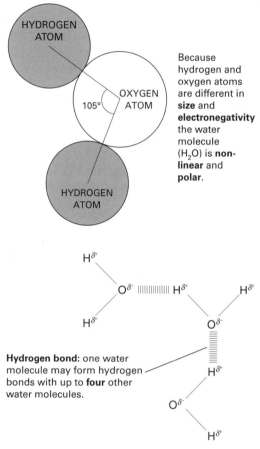

Because hydrogen and oxygen atoms are different in **size** and **electronegativity** the water molecule (H_2O) is **non-linear** and **polar**.

Hydrogen bond: one water molecule may form hydrogen bonds with up to **four** other water molecules.

This polarity means that individual water molecules can form **hydrogen bonds** with other water molecules. Although these individual hydrogen bonds are weak, collectively **they make water a much more stable substance than would otherwise be the case.**

- **High specific heat capacity:** the specific heat capacity of water (the amount of heat, measured in joules, required to raise 1 kg of water through 1°C) is very high: much of the heat absorbed is used to break the hydrogen bonds which hold the water molecules together.
- **High latent heat of vaporisation:** hydrogen bonds attract molecules of liquid water to one another and make it difficult for the molecules to escape as vapour: thus a relatively high energy input is necessary to vaporise water and water has a much higher boiling point than other molecules of the same size.
- **Molecular mobility:** the weakness of individual hydrogen bonds means that individual water molecules continually jostle one another when in the liquid phase.
- **Cohesion and surface tension:** hydrogen bonding causes water molecules to 'stick together', and also to stick to other molecules - the phenomenon of **cohesion**. At the surface of a liquid the inwardly-acting cohesive forces produce a 'surface tension' as the molecules are particularly attracted to one another.
- **Density and freezing properties:** as water cools towards its freezing point the individual molecules slow down sufficiently for each one to form its maximum number of hydrogen bonds. To do this the water molecules in liquid water must move further apart to give enough space for all four hydrogen bonds to fit into. As a result water expands as it freezes, so that ice is less dense than liquid water and therefore floats upon its surface.
- **Colloid formation:** some molecules have strong intramolecular forces which prevent their solution in water, but have charged surfaces which attract a covering of water molecules. This covering ensures that the molecules remain dispersed throughout the water, rather than forming large aggregates which could settle out. The dispersed particles and the liquid around them collectively form a **colloid**.

The biological importance of water

All living things depend on water, and its biological importance depends on its physical properties.

Volatility/stability: is balanced at Earth's temperatures so that a water cycle of evaporation, transpiration and precipitation is maintained.

Lubricant properties: water's cohesive and adhesive properties mean that it is viscous, making it a useful lubricant in biological systems.
For example:
- **synovial fluid** – lubricates many vertebrate joints;
- **pleural fluid** – minimises friction between lungs and thoracic cage (ribs) during breathing;
- **mucus** – permits easy passage of faeces down the colon, and lubricates the penis and vagina during intercourse.

Thermoregulation: the high specific heat capacity of water means that bodies composed largely of water (cells are typically 70–80% water) are very thermostable, and thus less prone to heat damage by changes in environmental temperatures.
The high latent heat of vaporisation of water means that a body can be considerably cooled with a minimal loss of water – this phenomenon is used extensively by mammals (sweating) and reptiles (gaping) and may be important in cooling transpiring leaves.

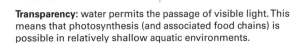

Supporting role: the cohesive forces between water molecules mean that it is not easily compressed, and thus it is an excellent medium for support. Important biological examples include the hydrostatic skeleton (e.g. earthworm), turgor pressure (in herbaceous parts of plants), amniotic fluid (which supports and protects the mammalian fetus) and as a general supporting medium (particularly for large aquatic mammals such as whales).

Molecular mobility: the rather weak nature of individual hydrogen bonds means that water molecules can move easily relative to one another – this allows osmosis (vital for uptake and movement of water) to take place.

Transpiration stream: the continuous column of water is able to move up the xylem because of cohesion between water molecules and adhesion between water and the walls of the xylem vessels.

Transparency: water permits the passage of visible light. This means that photosynthesis (and associated food chains) is possible in relatively shallow aquatic environments.

Expansion on freezing: since ice floats it forms at the surface of ponds and lakes - it therefore insulates organisms in the water below it, and allows the ice to thaw rapidly when temperatures rise. Changes in density also maintain circulation in large bodies of water, thus helping nutrient cycling. Floating ice also means that penguins and polar bears have somewhere to stand!

Solvent properties: allow water to act as a transport medium for polar solutes.
For example:
- movements of minerals to lakes and seas;
- transport via blood and lymph in multicellular animals;
- removal of metabolic wastes such as urea and ammonia in urine.

Metabolic functions
Water is used directly …
- as a reagent (source of reducing power) in photosynthesis;
- to hydrolyse macromolecules to their subunits, in digestion for example;
… and is also the medium in which all biochemical reactions take place.

Amino acids – the building blocks of proteins

Proteins are compounds containing the elements carbon, hydrogen, oxygen, and nitrogen. Some also contain sulfur.

Amino acids are compounds which are the building blocks from which proteins are made. There are 20 different amino acids.

$$\text{amino group } H_2N - \overset{\overset{\displaystyle R \text{ side group}}{|}}{\underset{\underset{\displaystyle H}{|}}{C}} - COOH \text{ carboxyl group}$$

The general structure of an amino acid

Each amino acid has a different R group. For example:

Amino acid	R group
glycine	$-H$
alanine	$-CH_3$
valine	$-CH(CH_2)_2$

There are 20 different R groups which is why there are 20 different amino acids.

Peptide bonds

A **dipeptide** is formed when two amino acid molecules combine.

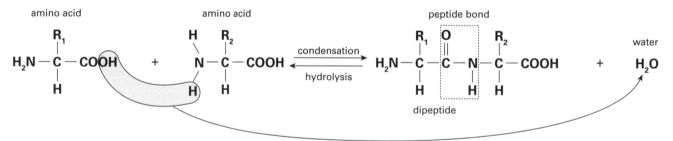

The structure of a dipeptide

Notice that

- the reaction is a **condensation** – the reverse **hydrolysis** would break the dipeptide into its component amino acid molecules
- the carboxyl group of one amino acid molecule reacts with the amino group of the other amino acid molecule – a molecule of water is eliminated
  - As a result a **peptide bond** forms between the two amino acid molecules.
- the type of dipeptide formed depends on the structures of R_1 and R_2

The more amino acid units joined together by condensation reactions, the larger is the resulting polymer.

Type of compound	Number of amino acid units joined together
dipeptide	2
peptide	3–20
polypeptide	21–50
protein	>50

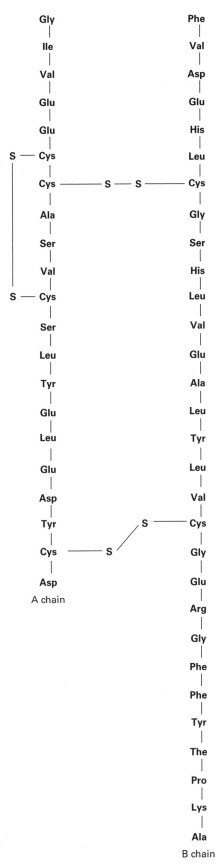

The primary structure of insulin

A chain (left column): Gly, Ile, Val, Glu, Glu, S—Cys, Cys, Ala, Ser, Val, S—Cys, Ser, Leu, Tyr, Glu, Leu, Glu, Asp, Tyr, Cys—S, Asp

B chain (right column): Phe, Val, Asp, Glu, His, Leu, Cys, Gly, Ser, His, Leu, Val, Glu, Ala, Leu, Tyr, Leu, Val, Cys, Gly, Glu, Arg, Gly, Phe, Phe, Tyr, The, Pro, Lys, Ala

You do not need to learn this structure but it gives you an idea of the sequence.

Protein structure

Primary (1°) structure

The order in which one amino acid unit follows another in the polypeptide chain(s) is unique for each protein. This unique amino acid sequence dictates the structure, chemical properties, and function of the particular protein. The sequence of amino acid units is called the **primary (1°) structure** of the protein. The diagram shows the primary structure of the hormone insulin. Notice that

- the insulin molecule consists of two polypeptide chains joined together
- one end of each polypeptide chain ends in an $-NH_3^+$ group (the amino end)
- the other end of each polypeptide chain ends in a $-COO^-$ group (the carboxyl end)

Secondary, **tertiary**, and **quaternary** structures of proteins arise from the primary structure of polypeptides.

The primary structure of a polypeptide chain allows

- hydrogen bonds to form between different amino acids along the chain
- interactions between R groups of the amino acids along the chain

  As a result the polypeptide chain bends and twists giving rise to the secondary, tertiary, and quaternary structures that shape the protein molecule.

Secondary (2°) structure

Secondary structure arises because of hydrogen bonding between the oxygen of the >C=O group of one amino acid unit and the hydrogen of the >NH group of another amino acid unit.

- If this bonding occurs within one polypeptide chain, the chain coils into an alpha helix (α-**helix**).
- If this bonding occurs between different, parallel polypeptide chains, the chains fold into a beta pleated sheet (β-**pleated sheet**).

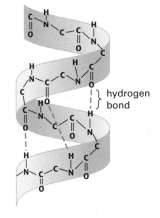

A polypeptide chain coiled into an α-helix

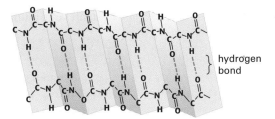

A polypeptide chain forming a β-pleated sheet

Tertiary (3°) structure

Tertiary structure arises when the α-helices and β-pleated sheets of many proteins fold and coil into a shape which is held in place by chemical bonds between different groups in the polypeptide chain.

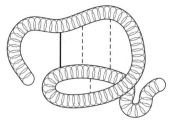

key

| = strong bond

┊ = weak bond

Different bonds hold together the tertiary structure of a protein

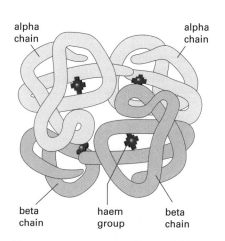

The quaternary structure of haemoglobin. Each alpha (α) and beta (β) polypeptide chain consists of about 140 amino acids.

> **Fact file**
>
> The weak bonds that hold the tertiary structure of a protein together include
> - disulfide (S–S) bonds
> - ionic interactions – attractions between positively charged and negatively charged areas of the molecule
> - hydrophobic and hydrophilic interactions – in water, hydrophobic parts of the polypeptide chain will be repelled and so be orientated towards the centre of the protein molecule. In contrast, hydrophilic areas will be attracted to the water and point outwards.

Quaternary (4°) structure

Quaternary structure arises when a protein molecule consists of two or more polypeptide chains. Haemoglobin (the oxygen carrying pigment in red blood cells) is an example. The way in which the chains fit together is maintained by the same types of chemical bond that hold together its tertiary structures.

Functions of proteins

The structure of proteins is a clue to their function in organisms.

Fibrous proteins consist of long polypeptide chains which run parallel to one another. Their function in cells and organisms is usually structural. **Collagen** is an example. Its fibres are insoluble, inelastic, and have high tensile strength. This makes collagen an ideal material for **tendons** and ligaments which must withstand large pulling forces.

> **Fact file**
>
> About 25% of the protein that makes the human body is collagen. It is a component of skin, teeth, bone, cartilage, and blood vessels as well as tendons and ligaments.

Globular proteins consist of polypeptide chains that fold up into a spherical shape. The environment inside cells and organisms is aqueous. The folding of globular proteins places the
- hydrophobic side chains in the interior of the molecule
- hydrophilic side chains on the outside of the molecule

   As a result water molecules are excluded from the hydrophobic centre of the folded protein molecule but gather around the hydrophilic surface.

   As a result globular proteins are usually soluble.

Enzymes, antibodies, and some hormones are globulins. Haemoglobin is another.

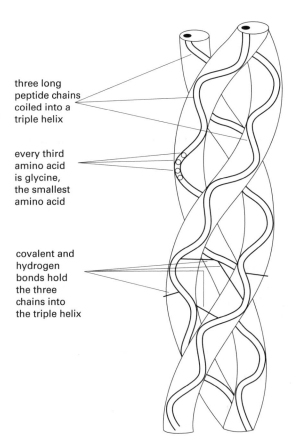

three long peptide chains coiled into a triple helix

every third amino acid is glycine, the smallest amino acid

covalent and hydrogen bonds hold the three chains into the triple helix

Collagen is the most abundant of all animal proteins. It is found in the connective tissue of skin, tendons and ligaments.
- the long fibres provide a 'framework' for tissues (like iron in reinforced concrete);
- presence of glycine allows three chains to pack together, giving strength to the molecule;
- side chains of other amino acids are hydrophobic, so the molecule is insoluble in water.

Testing for protein

Adding an alkaline solution of copper(II) sulfate to material that contains peptides or proteins produces a **pink to purple** colour. The test is called the **biuret** test. It detects peptide bonds. All peptides and proteins, therefore, give a positive result.

The biuret test is a **qualitative** test but can also be used semi-quantitatively. The colours produced depend on the number of peptide bonds in the test substance. Their range is:

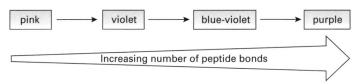

pink → violet → blue-violet → purple

Increasing number of peptide bonds

Proteins give a characteristic purple colour with the biuret test because the components contain the most peptide bonds per molecule.

Questions

1 Which elements make up a molecule of amino acid?

2 What is the primary structure of a protein?

3 What type of chemical reaction forms peptide bonds?

4 Briefly describe the biuret test for proteins.

2.03 Carbohydrates

Carbohydrates are compounds containing the elements carbon, hydrogen, and oxygen only. There are three main categories:

- **Monosaccharides** – single sugars. **Glucose** and **fructose** are examples. They are the basic molecular units (**monomers**) of which carbohydrates are made.
- **Disaccharides** – double sugars. **Maltose**, **sucrose**, and **lactose** are examples. One molecule of a disaccharide is formed when two monosaccharide molecules combine.
- **Polysaccharides** – compounds which are formed from the combination of hundreds of monosaccharide molecules. **Starch**, **glycogen**, and **cellulose** are examples.

Monosaccharides

In general the **molecular** formula of monosaccharides is written as $(CH_2O)_n$. Some examples are given in the table. Notice that the ratio of hydrogen and oxygen atoms is 2:1 – the same as water. The term carbohydrate means 'hydrated carbon'.

Value of n	Type of monosaccharide	Molecular formula	Examples
3	trioses	$C_3H_6O_3$	glyceraldehyde
5	pentoses	$C_5H_{10}O_5$	ribose
6	hexoses	$C_6H_{12}O_6$	glucose, fructose

Notice that glucose and fructose have the same molecular formula because they consist of the same atoms in the same proportions. However, their **structural** formulae show how the arrangement of the atoms of each molecule is different. We say that glucose and fructose are structural **isomers**.

Glucose molecules exist in two forms (**isomers**): (alpha) α-glucose and (beta) β-glucose. The diagram on the left shows α-glucose. If you imagine the molecules as 3D structures, with the hexagonal ring like a flat plate, the hydroxyl group (–OH) on carbon atom 1 is

- below the plane of the ring in α-glucose
- above the plane of the ring in β-glucose

Using 'shorthand' forms of molecules makes it easier to understand the structural changes taking place during chemical reactions.

Dissacharides

A **disaccharide** is formed when two monosaccharide molecules combine. For example, the diagram below shows how two molecules of α-glucose combine to form one molecule of **maltose**.

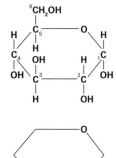

shorthand version

The structural formula of α-glucose. The numbers show the positions of the carbon atoms in the molecule.

shorthand version

The structural formula of fructose. The numbers show the positions of the carbon atoms in the molecule.

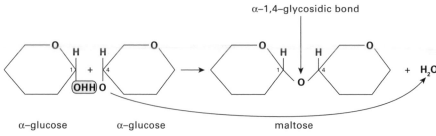

The formation of maltose

Notice that

- the reaction is a **condensation**. (The reverse **hydrolysis** breaks down maltose into its component α-glucose molecules.)
- the formation of a water molecule results in an oxygen 'bridge' joining the two molecules. The link is called a **glycosidic bond**.

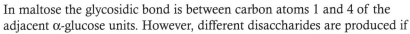

In maltose the glycosidic bond is between carbon atoms 1 and 4 of the adjacent α-glucose units. However, different disaccharides are produced if

- the combination of monosaccharide molecules is different and
- the links between carbon atoms in adjacent monosaccharide molecules is different

For example:

- **sucrose** = α-glucose + β-fructose
- **lactose** = α-glucose + galactose

Polysaccharides

Polysaccharides are formed from many monosaccharide molecules joined by condensation reactions. Their general formula is:

$$nC_6H_{12}O_6 \rightarrow (C_6H_{10}O_5)_n + nH_2O \qquad \text{where } n = 10^2 \rightarrow 10^3$$

For example, **starch** is made from many condensation reactions forming glycosidic bonds between α-glucose molecules.

Testing for sugars

All monosaccharides (e.g. glucose and fructose) and some disaccharides (e.g. maltose and lactose, but not sucrose) are **reducing sugars**. They contain a free **aldehyde** group ($-CHO$) or **ketone** group ($>C=O$) which reduces copper(II) ions (Cu^{2+}) to copper(I) ions (Cu^+) when heated in an alkaline solution. This is the basis of the **Benedict's test** for reducing sugars.

- Benedict's reagent turns from blue to red when heated with a reducing sugar.

Testing for non-reducing sugars

Sucrose is an example of a **non-reducing sugar**. It does not give a positive result to a simple Benedict's test. However, if sucrose solution is first heated with dilute hydrochloric acid then it hydrolyses into its component monosaccharide molecules. These then test positive when heated with Benedict's reagent. This is the basis of the test for a non-reducing sugar.

- The sugar solution gives a negative result when heated with Benedict's solution (no colour change).
- When hydrolysed (by heating with acid) and then neutralized (by adding sodium hydrogencarbonate), testing again gives a positive result to the Benedict's test (changes from blue to red).

Testing for glucose using colorimetry

Light is absorbed as it passes through a solution of a coloured substance. The amount of light absorbed by the solution depends on the concentration of the coloured substance.

Benedict's test produces a range of colours depending on the concentration of reducing sugar in the solution being tested. The colorimeter can provide a quantitative result from this test.

The food is tested for glucose by boiling with Benedict's reagent. The resulting mixture is cooled and filtered. The absorbance of the supernatant that remains after filtering is then measured in a colorimeter using a red filter. The greater the absorbance, the lower the glucose concentration. The absorbance can be compared with a calibration curve to give a quantifiable concentration of glucose.

Testing for starch

Starch doesn't dissolve in water but forms a suspension. Iodine doesn't dissolve in water either, but does dissolve in potassium iodide solution. When this iodine solution is added to the starch suspension, it turns from yellow-orange to an intense blue-black. This is the basis of the **iodine test** for starch.

- Starch turns iodine solution blue-black.

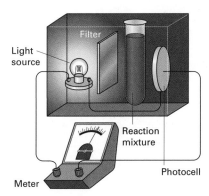

A colorimeter measures the concentration of a coloured substance

Questions

1 The term 'carbohydrate' means hydrated carbon. Explain the link between the term and the general formula for a monosaccharide.

2 Why are molecules of glucose and fructose described as structural isomers?

3 Briefly describe a test which identifies the presence of non-reducing sugars.

2.04 Starch, glycogen, and cellulose

Alpha and beta glucose

Recall that the –OH (hydroxyl) group on carbon atom 1 of an α glucose molecule is *below* the plane of the ring of carbon atoms. In β glucose it is *above* the plane.

α glucose   β glucose

Starch

Starch is a polysaccharide formed from condensation reactions between many molecules of α glucose. It has two components:

- **amylose** – consists of unbranched chains of α 1–4 glycosidic bonds between α glucose molecules. Each chain coils like a spiral staircase into a helical structure.

α 1–4 linkage

glucose units

amylose

- **amylopectin** – consists of branched chains with 1–6 glycosidic bonds between α glucose molecules forming the branches

α 1–6 linkage

α 1–4 linkage

glucose units

side branch

main branch

amylopectin

The helical shape of amylose makes for a dense polysaccharide, enabling many subunits to be packed into a small space. Starch is an excellent storage molecule.

Glycogen

Glycogen is like amylopectin except that it consists of

- fewer 1–4 glycosidic bonds between α glucose molecules
- many more 1–6 linkages

The molecule is more branched than amylopectin, making glycogen less dense and more soluble than starch.

Cellulose

Cellulose is formed from condensation reactions between many molecules of β glucose.

- The –OH (hydroxyl) group on carbon atom 1 is *above* the plane of the ring of carbon atoms.
- The –OH group on carbon atom 4 is *below* the plane of the ring.
  - 🐝 As a result, for the –OH groups on carbon atoms 1 and 4 of adjacent β glucose molecules to form a glycosidic bond, one molecule has to be flipped over (rotated through 180°) relative to its next-door neighbour:

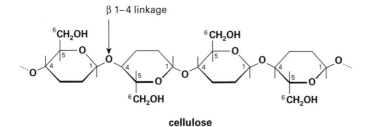

cellulose

The formation of β 1–4 glycosidic linkages produces rigid chain-like molecules.

- **Hydrogen bonds** cross-link chains into bundles called **microfibrils**.
- Further hydrogen bonding binds microfibrils into **fibres**.

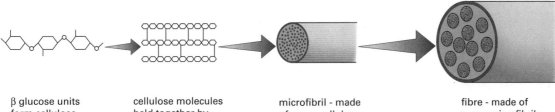

| β glucose units form cellulose molecules | cellulose molecules held together by hydrogen bonds | microfibril - made of many cellulose molecules | fibre - made of many microfibrils |

Cellulose has a tightly bundled structure.

Carbohydrates in living organisms

Respiratory substrate

Glucose is a small soluble molecule which makes it easy to transport into and out of cells. It polymerises to form storage polysaccharides which makes it a useful molecule in living things.

Food stores

Starch and glycogen are almost insoluble which means that they have little effect on the osmotic properties of cells. Therefore they can be stored in cells as energy reserves.

- Starch is stored in plant cells.
- Glycogen is stored in animal cells.

Enzyme-catalysed reactions break down starch and glycogen into glucose. When glucose is oxidized, energy is released and carbon dioxide and water formed. The term **cellular respiration** refers to the series of reactions.

Structural materials

Cellulose fibres have high tensile strength which means that each one is difficult to pull apart. Up to 40% of the wall of a plant cell is made of cellulose. The arrangement of cellulose fibres gives shape to the cell as it grows.

The mechanical strength of the fibres also means that plant cells can withstand the large hydrostatic pressures that develop inside them without bursting as the result of osmosis.

Questions

1. How do the structures of α glucose and β glucose molecules differ?

2. Briefly explain why plant cells do not burst when large hydrostatic pressures develop inside them as the result of osmosis.

3. Why do starch and glycogen have little effect on the osmotic properties of cells?

2.05 Lipids

Lipids

Lipids are compounds containing the elements carbon, hydrogen, and oxygen only. **Fats** and **oils** produced by plants and animals are lipids – fats are solid at room temperature, oils are liquid at room temperature.

Most lipids are mixtures of **triglycerides**.

- A triglyceride is an ester of fatty acids and glycerol (an ester is a substance formed from the reaction between an acid and an alcohol).
- A molecule of fatty acid consists of a hydrocarbon chain and a carboxyl group ($-COOH$). The table gives examples.

Fatty acid	Molecular formula
stearic acid	$C_{17}H_{35}COOH$
oleic acid	$C_{17}H_{33}COOH$
palmitic acid	$C_{15}H_{31}COOH$

The formula of glycerol is:

$$CH_2-OH$$
$$|$$
$$CH-OH$$
$$|$$
$$CH_2-OH$$

Molecules of fatty acid and glycerol combine to form a triglyceride like this:

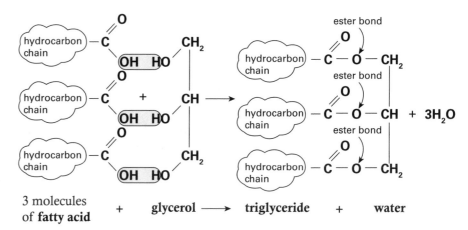

3 molecules of **fatty acid** + glycerol ⟶ triglyceride + water

Notice that

- the combination of molecules of fatty acid with the molecule of glycerol is a condensation reaction
- three molecules of fatty acid are combined with one molecule of glycerol producing a triglyceride and three molecules of water
  - a **simple** triglyceride is formed if the fatty acid molecules are the same
  - a **mixed** triglyceride is formed if the fatty acid molecules are different

Saturation and unsaturation

Saturated – all the bonds linking adjacent carbon atoms of the hydrocarbon chain of the fatty acid are single bonds:

$$CH_3-CH_2-CH_2-CH_2-CH_2-CH_2-CH_2-CH_2-$$

Monounsaturated – the hydrocarbon chain of the fatty acid contains one double bond:

$$CH_3-CH_2-CH=CH-CH_2-CH_2-CH_2-CH_2-$$

Polyunsaturated – the hydrocarbon chain of the fatty acid contains more than one double bond:

$$CH_3-CH_2-CH=CH-CH_2-CH=CH-CH_2-$$

- Saturated fatty acids combine with glycerol to form saturated triglycerides.
- Unsaturated fatty acids combine with glycerol to form unsaturated triglycerides.

Remember that the valency of a carbon atom is 4.

Fact file

The structure of triglycerides makes them useful in living organisms:

- They are insoluble in water and also store a large amount of chemical energy, making fats and oils useful storage molecules.
- They are readily hydrolysed to fatty acids and glycerol. The fatty acids can be used for respiration in cells when glucose is not available.
- They dissolve in organic solvents so can pass through the phospholipid bilayer of cell membranes.

The emulsion test for lipids

Different tests may be used to detect the presence of lipids. One of them is the **emulsion test**:

- Mix ethanol and the test material in equal volumes. Shake to help dissolve any lipids present.
- Then add an equal volume of water and shake the mixture again. A milky white emulsion indicates the presence of lipids.

Phospholipids

Phospholipids are triglycerides which contain a phosphate group instead of one of the fatty acid components.

- The phosphate group dissolves in water and is therefore **hydrophilic** (literally 'water loving').
- The hydrocarbon chains of the two fatty acid components do not dissolve in water and are therefore **hydrophobic** (literally 'water hating').

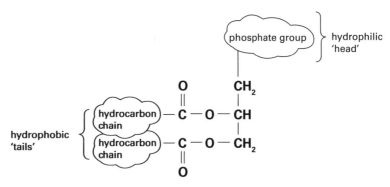

The structure and function of cell membranes depend on the hydrophilic and hydrophobic properties of phospholipid molecules.

Remember the functions of phospholipids and cholesterol in plasma membranes

- **Phosopholipids** are an important part of the plasma membrane. Remember that they have hydrophilic heads (phosphate groups) and hydrophobic tails (hydrocarbon chains). Remember also that cell contents are aqueous and cells are bathed in water.
    - As a result, in phospholipid molecules spontaneously form a stable two layer framework called a phospholipid **bilayer**. The hydrophobic tails point inwards, shielded from the water. The hydrophilic heads face outwards, forming hydrogen bonds with the water.
- The hydrocarbon tails of many of the phospholipid molecules are 'kinked' because of the presence of double bonds.
    - As a result phospholipids are loosely packed together.
    - As a result the plasma membrane is fluid.
    - As a result proteins and other material can move sideways within the membrane.
- Cholesterol is a lipid. Its molecules, wedged into the bilayer, help to keep the plasma membrane fluid at low temperatures.

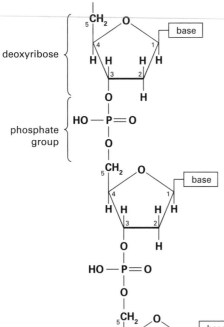

deoxyribose

phosphate group

A strand of DNA

Deoxyribonucleic acid (DNA) is a nucleic acid. Its molecules contain the genetic information which instructs cells to synthesize (make) **ribonucleic acid (RNA** – another type of nucleic acid) and proteins.

'Genetic' means that the information is inherited by (passed on to) daughter cells when parent cells divide.

- The sections of DNA that carry genetic information are called **genes**.
- Other sections of DNA control the use of this genetic information.

Nucleotides

DNA is a polynucleotide. This means it is a polymer made up of monomers called **nucleotides**. A nucleotide in the case of DNA has three components – a molecule each of

- the pentose sugar **deoxyribose**
- a **base** which is either **adenine (A)**, **thymine (T)**, **guanine (G)**, or **cytosine (C)**
- a **phosphate** group

Condensation reactions join the components together forming a nucleotide.

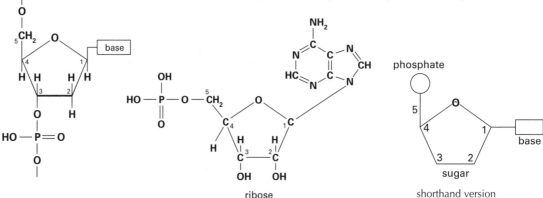

ribose

shorthand version

Structure of a nucleotide

DNA structure

Many condensation reactions join together nucleotide units forming a strand of DNA.

Notice in the diagram above left:

- Adjacent sugar rings join through the phosphate group from carbon atom 3 of one sugar to carbon atom 5 of the next sugar in line.
- These links are **phosphodiester** bonds and hold the DNA strand together.

A shorthand way of describing this linkage is 3' → 5' → 3'.

A molecule of DNA is made of two polynucleotide strands. The strands coil round each other forming a '**double helix**'. This is stabilized by hydrogen bonds between the bases attached to the two strands.

Notice in the diagram on the left:

- A only bonds with T
- G only bonds with C

This arrangement is called **complementary base pairing**.

complementary base pairs

one complete turn of the helix is 3.4 nm (= 10 base pairs)

sugar–phosphate backbone

key

S = deoxyribose sugar

P = phosphate sugar

A, T, G, C = different bases

The double helix structure of DNA

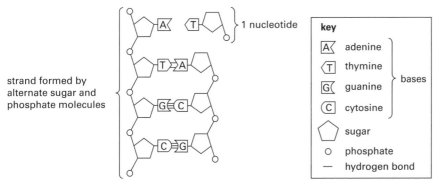

Complementary base pairing in DNA

key

A⟨	adenine
⟨T	thymine
G⟨	guanine
C	cytosine

bases

⬠	sugar
○	phosphate
—	hydrogen bond

strand formed by alternate sugar and phosphate molecules

Notice in the diagram above:
- one strand runs from carbon atom $3' \rightarrow 5' \rightarrow 3'$... and so on in one direction
- its partner strand runs $5' \rightarrow 3' \rightarrow 5'$ in the opposite direction

We say that the strands are **anti-parallel**.

The hydrogen bonds linking bases are sufficiently strong and numerous to hold together a molecule of nucleic acid. But they are also weak enough to break during DNA replication or when genetic information is being transcribed.

RNA structure

RNA is another polynucleotide. It differs from DNA in that:
- The sugar units in the chain are not deoxyribose as in DNA, but ribose (carbon number 2 has an –OH group).
- The base thymine is replaced by a different base called uracil. So the bases in RNA are adenine (A), thymine (T), cytosine (C), and guanine (G).
- RNA is usually single stranded rather than double stranded.

ribose

Structure of an RNA nucleotide

Types of RNA

Messenger RNA (mRNA) is a single stranded molecule which is formed alongside DNA in the nucleus by a process called **transcription**. mRNA has a sequence of bases complementary to the sequence in the DNA strand. It carries the genetic code from the nucleus to the cytoplasm where proteins are made.

Transfer RNA (tRNA) has a complex three-dimensional shape formed by hydrogen bonding within the molecule. It carries specific amino acids to the ribosomes during protein synthesis.

Ribosomal RNA (rRNA) is single stranded but folded into a complex series of shapes which combine with proteins to form the ribosomes. This is the site of protein synthesis.

Discovering the structure of DNA

By the 1950s, it was recognized that DNA is the genetic material of cells. Discovering its structure was a priority and different techniques were used to solve the problem. Success came in 1953 when James Watson and Francis Crick combined all the evidence from different sources. They succeeded in building a model of DNA which fitted all of the known facts.

At the time Rosalind Franklin was using X-ray crystallography to probe the arrangement of the atoms of DNA molecules. The technique involved focusing a beam of X-rays through a fibre of DNA on to a photographic plate.

- X-rays affect photographic emulsion. A pattern of dots can be seen when the plate is developed.
- The pattern of dots represents the position of the atoms making up the molecule.
- Plotting the density of the dots (and therefore of the atoms) and their spatial relationships allows the information in the pattern to be interpreted as a 3D model of the molecule.

Franklin's work played a key role in solving the problem of the structure of DNA.

Questions

1 What does the term 'genetic' mean?

2 Briefly explain the relationship between phosphodiester bonds, complementary base pairing, and a molecule of double-stranded DNA.

3 What are the components of a nucleotide?

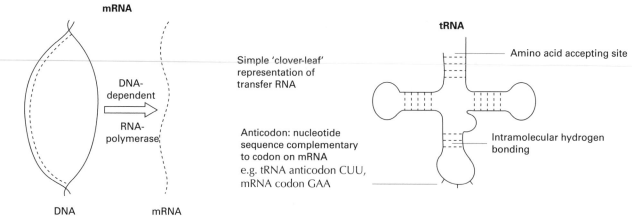

The structure of mRNA and tRNA

Roles of DNA and RNA in living organisms

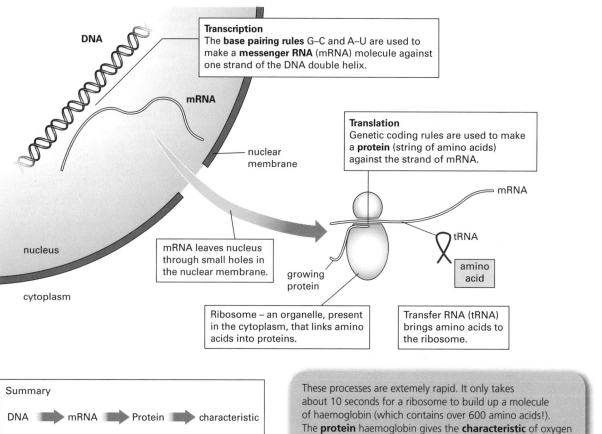

Transcription
The **base pairing rules** G–C and A–U are used to make a **messenger RNA** (mRNA) molecule against one strand of the DNA double helix.

Translation
Genetic coding rules are used to make a **protein** (string of amino acids) against the strand of mRNA.

nuclear membrane

mRNA leaves nucleus through small holes in the nuclear membrane.

Ribosome – an organelle, present in the cytoplasm, that links amino acids into proteins.

Transfer RNA (tRNA) brings amino acids to the ribosome.

growing protein

amino acid

tRNA

mRNA

nucleus

cytoplasm

Summary

DNA ➡ mRNA ➡ Protein ➡ characteristic

These processes are extemely rapid. It only takes about 10 seconds for a ribosome to build up a molecule of haemoglobin (which contains over 600 amino acids!). The **protein** haemoglobin gives the **characteristic** of oxygen transport to red blood cells.

How information in DNA codes for characteristics in cells

2.07 DNA replication

Making an exact copy

During **replication** DNA makes an exact copy of itself. The process is a necessary part of the division of a cell nucleus.

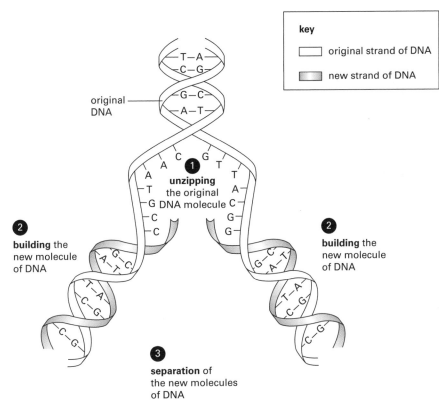

key

☐ original strand of DNA

▨ new strand of DNA

DNA replication – each strand of unzipped DNA is a pattern (template) against which a new strand forms.

There are three stages.

❶ Unzipping the original DNA molecule:
- The enzyme **DNA helicase** catalyses the breaking of hydrogen bonds linking the base pairs of the two strands of DNA.
- The double helix unzips as the base pairs separate.

❷ Building the new polynucleotide chain of DNA:
- Nucleotides free in solution within the nucleus each link with their complementary base on either of the unzipped strands of DNA (each called a template strand). Linkage is catalysed by the enzyme **DNA polymerase**.
- The process repeats itself again and again in the $5' \rightarrow 3' \rightarrow 5'$ direction along both strands of template DNA.
- As a result a new polynucleotide strand grows against each template strand.

❸ Separation of the new DNA molecules:
- When all of the bases of each template strand of DNA are each joined with the complementary base of a free nucleotide and these nucleotides have linked together, replication is complete.
- The two new molecules of DNA separate.

Semi-conservative replication

DNA replication occurs during interphase. The process is **semi-conservative** in that each new DNA molecule consists of a strand of template DNA (arising from the unzipping of the original double helix) and a strand of DNA formed as a complement of the template strand.

Questions

1 What are the roles of DNA helicase and DNA polymerase during DNA replication?

2 Why is DNA replication described as semi-conservative?

3 What does the phrase 'DNA template strand' mean?

Genes

Some sections of strands of DNA carry information which enables cells to synthesize (make) molecules of polypeptide. The information is carried in the sequence of bases on the nucleotides which make up these sections. The sections are called **genes**.

The differences between genes are the result of differences in the sequences of their bases.

Estimates suggest that most human cells each carry 20 000 to 30 000 genes, enabling the cells to synthesize millions of different types of polypeptide.

Genes are often categorized as:

- **structural genes** – affecting the synthesis of enzymes and the other polypeptides that make up body structures. For example collagen accounts for up to 25% of total body protein (tendons, ligaments, connective tissue, etc.)
- **regulatory genes** – affecting the synthesis of polypeptides which control the development of organism. Regulatory genes may also affect the activity of other genes.

The term **locus** refers to the position of a gene on a particular strand of DNA. The ordered list of gene loci (plural of locus) is called a **genetic map**.

The genetic code

To make a molecule of a particular polypeptide, many amino acid units must combine in the correct order. The sequence of bases of a gene is responsible for getting the order correct. In other words, the sequence carries the information which enables a cell to assemble that particular polypeptide.

The sequence of bases of all of the genes of a cell, and the information each sequence carries, is the **genetic code**. The code is almost universal. It is the same in the cells of most living things.

The information needed to assemble one amino acid unit in its correct place in a polypeptide is contained in a sequence of three bases. The sequence is called a **codon**. A gene therefore is a sequence of codons and the genetic code is all of the codons in the DNA of a cell.

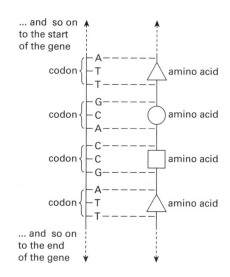

Each triplet code in a gene forms a codon for a particular amino acid. The sequence of codons controls the sequence of amino acids in a polypeptide.

Coding and non-coding DNA

In eukaryotes most genes are not a continuous sequence of codons. Non-coding regions (called **introns**) split up the coding regions (called **exons**).

Remember that a strand of DNA may consist of thousands of nucleotides joined together. Each nucleotide has a base (A, T, C, or G) as part of its structure. A strand of DNA, therefore, carries a sequence (a particular order) of bases.

Fact file

We say that the genetic code is **degenerate** because nearly all of the amino acids are each coded for by more than one codon.

Of the 64 possible codons:

- three are stop codons – during protein synthesis 'stop' means 'end of polypeptide chain'
- one codon is an 'initiator' – during protein synthesis this means 'start of polypeptide chain'

Questions

1 What is a gene?
2 Why is the genetic code described as a triplet code?
3 What does the term 'locus' mean?

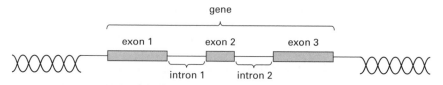

A gene is usually a sequence of exons and introns.

Also there are non-coding regions of DNA between genes. These non-coding regions often consist of short sequences of bases which repeat over and over again (multiple repeats) and are called **mini-satellites**.

2.09 Enzymes

Most enzymes are globular proteins. They are **catalysts**. In general, a catalyst
- alters the rate of a chemical reaction
- is effective in small amounts
- is involved in but unchanged by the chemical reaction it catalyses

All the features of catalysts are also features of enzymes. In addition enzymes are
- **specific** in their action, catalysing a particular chemical reaction or type of reaction
- sensitive to changes in pH and temperature

> **Remember**
> - The substance an enzyme helps to react is the **substrate**.
> - The substance formed by the reaction is the **product**.
> - When an enzyme and its substrate bind together an **enzyme–substrate complex** is formed.
> - Within the enzyme–substrate complex, the substrate undergoes reaction forming an **enzyme–product complex**.
> - The products then leave the active site, liberating the enzyme to catalyse another reaction.

Only a small part of an enzyme molecule binds with its substrate molecule. The part is called the **active site**. It consists of just a few of the amino acid units that make up the enzyme molecule as a whole.

Enzyme action may be **intracellular**, catalysing reactions inside the cell where the enzyme is made. Alternatively in **extracellular** action the enzyme may be secreted outside the cell and catalyse reactions outside the cell.

Enzymes lower activation energy

Activation energy is the amount of energy required to bring about any particular chemical reaction. Without it chemical reactions will not take place.

Enzymes lower activation energy by forming **enzyme–substrate complexes**. They enable reactions which would need high temperatures in the laboratory to take place at body temperature.

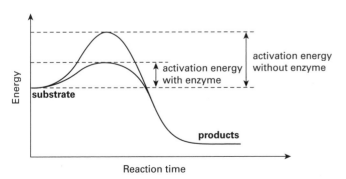

How do enzymes work?

The shape of an enzyme, like all proteins, is the result of its tertiary structure. An enzyme's active site also has a precise shape. An enzyme will bind with a particular substrate molecule because the shape of the active site is **complementary** to (opposite to) the shape of the substrate molecule. The two shapes fit like a key fits into a lock.

The idea of **lock and key** helps to explain why a particular enzyme will only catalyse a particular chemical reaction (or type of reaction). Only the shape of the substrate molecule in question fits the active site of the enzyme. The diagram shows an example of this.

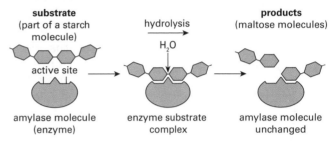

The enzyme amylase binds with starch, catalysing the breaking of alternate glycosidic bonds. The reactions are hydrolyses and maltose is formed.

> 'Lock and key' suggests that the active site of an enzyme and its substrate are *exactly* complementary. Recent work favours the **induced fit hypothesis**:
> - The active site and substrate are fully complementary only *after* binding has taken place.
> - The initial binding of a substrate molecule to the active site alters the shape (tertiary structure) of the active site.
>   - As a result the shape of the substrate molecule then alters, assisting the reaction to take place.

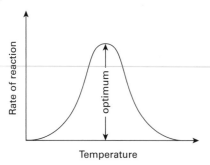

Enzyme activity varies with temperature

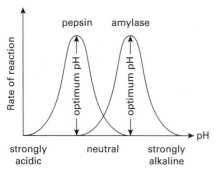

Enzyme activity varies with pH

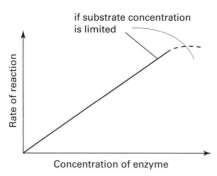

Enzyme activity varies with the concentration of enzyme

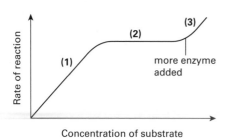

Enzyme activity varies with the concentration of substrate

Factors affecting the activity of enzymes

Remember that any factor that affects the shape of an enzyme molecule, in particular its active site, affects the activity of the enzyme.

Temperature

In general, the rate of chemical reactions increases with temperature (doubles for every 10 °C rise in temperature). But for enzyme-catalysed reactions, this is true only within a limited range of temperature. Once an enzyme reaches its optimum, any further increase in temperature causes a decrease in the rate of reaction.

The decrease is caused by a permanent change in the shape of the enzyme (and its active site). We say that the enzyme is **denatured**. Its active site is no longer complementary with the substrate molecule.

Qs and As

Q What does 'optimum' mean when we talk about the effect of temperature on enzyme activity?

A *The 'optimum' is the temperature value at which the number of collisions between an active enzyme and its substrate is at a maximum. As a result the rate of the enzyme-catalysed reaction is at a maximum.*

---

pH

A change of pH from the optimum for a particular enzyme alters the electric charge carried by the amino acid units forming the active site. The enzyme (and its active site) is denatured, and is no longer complementary with the substrate molecule. This causes a decrease in the rate of reaction.

Concentration of enzyme

Enzymes are not used up during catalysis, and enzymes can be used over and over again; enzymes therefore work very well at low concentrations. Increasing the enzyme concentration provides more active sites so the rate of enzyme activity increases as long as the substrate is present in excess.

Concentration of substrate

An increase in the concentration of substrate affects the rate of reaction for a fixed concentration of enzyme.

Notice that

- if there is an excess of enzyme, the rate of reaction is directly proportional to the concentration of the substrate **(1)**
- when all the enzyme active sites are occupied by substrate molecules the rate of reaction is limited **(2)**
- addition of more enzyme results in an increase in the rate of reaction which is proportional to the concentration of the substrate **(3)**

The effects of substrate concentration on the rate of reaction assume that all the other conditions that affect the rate of enzyme-catalysed reactions (e.g. pH, temperature) are constant.

Questions

1 What is activation energy and how do enzymes affect it?
2 Use the induced fit hypothesis to describe the binding of an enzyme with its substrate.
3 Why does denaturation affect the activity of an enzyme?

2.10 Experiments on enzyme action

Investigating factors

The rate of an enzyme-catalysed reaction can be investigated in the laboratory to see how the rate of reaction is affected by factors such as

- temperature
- pH
- enzyme concentration
- substrate concentration.

One commonly studied reaction is that of the enzyme catalase on hydrogen peroxide. Catalase catalyses the breakdown of hydrogen peroxide into oxygen and water. The volume of oxygen evolved can be measured to give a measure of reaction rate.

Catalase is present in potato tissue. Discs of potato are put into a solution of hydrogen peroxide. The catalase breaks down the hydrogen peroxide into water and oxygen, and the oxygen passes to the manometer. The changing level of manometer fluid shows how much oxygen is being produced, giving a measure of activity of the enzyme. The diagram shows the practical setup.

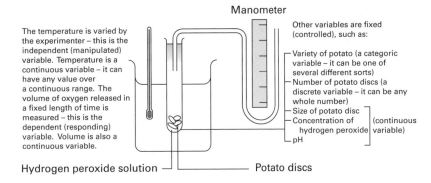

An identical control experiment is also set up, in which the manipulated variable (e.g. the temperature) is not changed. This confirms that the changes in temperature, and not some unknown variable, are causing any observed changes in enzyme activity.

The rate of oxygen evolution can be measured while varying a factor under investigation such as the temperature, the amount of enzyme (number or size of potato discs), concentration of substrate (hydrogen peroxide), or pH.

Recording and manipulating data

The raw results are collected in a table. An example is shown below. The first two columns show **raw data**, collected by the experimenter. The third column is **manipulated data** – it is calculated by the experimenter from the first two columns.

Temperature (°C)	Time taken to produce 10 cm³ of oxygen (s)	Rate of oxygen release (cm³/s)
15	40	0.3
25	20	0.5
35	5	1.2
45	20	0.5
55	40	0.3
65	120	0.1
75	no gas produced	0.0

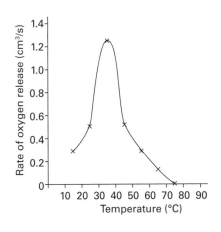

A graph plotted of rate of oxygen release against temperature gives the classic curve of enzyme activity.

2.11 Enzyme inhibition

Competitive inhibition

A **competitive inhibitor** is a substance that combines with the active site of an enzyme, preventing its normal substrate from binding with it. The diagram shows you the idea.

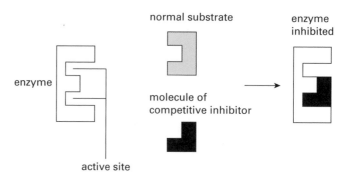

Enzyme inhibition

Non-competitive inhibition

A **non-competitive inhibitor** is a substance that combines with some part of an enzyme molecule *other* than its active site. The change in the shape of the enzyme molecule causes a change in the shape of its active site. The substrate molecule is no longer able to bind to the active site.

The inhibition may be

- **reversible** – breaking up the inhibitor–enzyme complex *is* possible
- **irreversible** – breaking up the inhibitor–enzyme complex *is not* possible. Heavy metals such as arsenic (As) and mercury (Hg) are irreversible non-competitive inhibitors.

Allosteric inhibition

In some enzymes, a chemical group other than the active site can bind with a substance other than the normal substrate. The group is called the **allosteric site**. An **allosteric inhibitor** is a substance which reversibly combines with the allosteric site.

Cofactors and coenzymes

Some enzymes have one or more non-protein parts to the molecule that are needed for the enzyme to function. These may be ions or other non-protein molecules, and are called **cofactors**. The inactive enzyme without its cofactor is an **apoenzyme**, and when combined with its cofactor in the active form it becomes a **holoenzyme**.

A **coenzyme** is a cofactor that is a complex non-organic molecule that is loosely associated with the enzyme molecule. Many vitamins are coenzymes. Cofactors play a central role in metabolism, regulating biochemical pathways by switching enzymes on and off.

Medicines and poisons

The reversible inhibition of enzymes plays a central role in medicine, and many metabolic poisons act by irreversibly inhibiting en enzyme.

- Aspirin reversibly inhibits the enzyme PGHS, involved in forming prostaglandins. This group of chemicals is involved in the inflammatory response and increases susceptibility to pain, so blocking the formation of prostaglandins has a painkilling effect.
- Cyanide is a respiratory poison. It irreversibly inhibits a respiratory enzyme called cytochrome oxidase.
- Some nerve gases irreversibly block the enzyme acetylcholinesterase. This enzyme normally breaks down the neurotransmitter acetylcholine as soon as it has acted. The continued presence of the neurotransmitter causes the muscles to go into spasm, preventing breathing and so causing death.

2.12 Diet and malnutrition

Balanced diet

A balanced human diet contains fat, protein, carbohydrate, vitamins, minerals, water and fibre **in the correct proportions**.

Carbohydrates

Mainly as a **respiratory substrate**, i.e. to be oxidized to release **energy** for active transport, synthesis of macromolecules, cell division and muscle contraction.

Common sources: rice, potatoes, wheat and other cereal grains, i.e. as **starch**, and as refined sugar, **sucrose**, in food sweetenings and preservatives.

Digested in duodenum and ileum and absorbed as **glucose**.

Lipids

Highly reduced and therefore can be oxidised to release **energy**. Also important in **cell membranes** and as a component of **steroid hormones**

Common sources: meat and animal foods are rich in **saturated fats** and **cholesterol**, plant sources such as sunflower and soya are rich in **unsaturated fats**.

Digested in duodenum and ileum and absorbed as **fatty acids and glycerol**.

Vitamins have no common structure or function but are essential in small amounts to use other dietary components efficiently. **Fat-soluble vitamins** (e.g. A, D and E) are ingested with fatty foods and **water-soluble vitamins** (B group, C) are common in fruits and vegetables.

Fibre (originally known as **roughage**) is mainly cellulose from plant cell walls and is common in fresh vegetables and cereals. It may provide some energy but mainly serves to aid faeces formation, prevent constipation and ensure the continued health of the muscles of large intestine.

Proteins are **building blocks** for growth and repair of many body tissues (e.g. myosin in muscle, collagen in connective tissues), as **enzymes**, as **transport systems** (e.g. haemoglobin), as **hormones** (e.g. insulin) and as **antibodies**.

Common source: meat, fish, eggs and legumes/pulses. Must contain eight **essential amino acids** since humans are not able to synthesise them. Animal sources generally contain more of the essential amino acids.

Protein quality can be assessed quantitatively as

Apparent digestibility of protein

$$= \frac{\text{N (nitrogen) intake} - \text{N in faeces}}{\text{N intake}}$$

and

Biological value of growth and maintenance

$$= \frac{\text{N intake} - \text{N in faeces} - \text{N in urine}}{\text{N intake} - \text{N in faeces}}$$

Digested in stomach, duodenum and ileum and absorbed as **amino acids**.

Minerals have a range of **specific** roles (direct structural components, e.g. Ca^{2+} in bones; constituents of macromolecules, e.g. PO_4^{3-} in DNA; part of pumping systems, e.g. Na^+ in glucose uptake; enzyme cofactors, e.g. Fe^{3+} in catalase; electron transfer, e.g. Cu^{2+} in cytochromes) and **collectively** help to maintain solute concentrations essential for control of water movement. They are usually ingested with other foods – dairy products and meats are particularly important sources.

Water is required as a solvent, a transport medium, a substrate in hydrolytic reactions and for lubrication. A human requires 2–3 dm^3 of water daily, most commonly from drinks and liquid foods.

An adequate diet provides sufficient **energy** for the performance of metabolic work, although the 'energy food' is in unspecified form.

A balanced diet provides all dietary requirements **in the correct proportions**. Ideally this would be $^1/_7$ fat, $^1/_7$ protein and $^5/_7$ carbohydrate.

Eating a diet that is not adequate or nor balanced leads to **malnutrition**.

Obesity

Obesity is one form of malnutrition caused by eating more than the body needs.

Men who are 20% overweight have a 25% increase in mortality – mainly from diabetes, coronary heart disease and cerebrovascular disorders.

Obese people also have increased risk of:

- hiatus hernia
- breast cancer
- endometrial cancer
- prostate enlargement
- menstrual abnormalities.

Arteries/arterioles: arteriosclerosis and hypertension → increased risk of stroke or brain haemorrhage.
Greater danger of thrombosis.

Heart: cardiac arrest is more likely (coronary artery occlusion and high blood pressure).

Liver: increased fatty degeneration of liver tissue.

Lungs: infections are more likely since greater body mass impairs movement of the diaphragm and therefore limits tidal flow in and out of lungs.

Vertebral column: greater body mass causes compression wear to vertebrae – scoliosis and sciatic nerve trap may result.

Pancreas: incidence of pancreatic cancer increases. Endocrine function impaired – obesity increases the risk of onset of diabetes and associated complications.

Hip joints: fracture of the head of the femur – particularly in obese, post-menopausal females.

Knee joints: degenerative disease more likely.

Gall bladder: Gallstones are more frequent – these are largely cholesterol stones.

Possible health problems associated with being overweight

Body mass index: are you obese?

Tables of desirable weights, with allowance for age, sex and height, may be consulted ~ 20% greater than ideal weight is considered **morbid obesity**.

$$\text{body mass index} = \frac{\text{mass in kg}}{(\text{height in metres})^2}$$

	Ideal	Morbid obesity
Male	20–25	30+
Female	18.5–23.5	28.5+

Treatment for obesity

- **Dietary control**
  Long term use of a balanced diet is likely to be successful: 'crash' diets are to be avoided.

- **Drug therapy**
  – may depress appetite with serotonin mimics
  – may promote metabolism with amphetamines.
    Long- term use is addictive.

- **Surgery**
  Jaw wiring limits food intake.
  Gastric stapling increases feeling of satiation.

The Energy Balance: Intake in the diet and **Demand** for metabolism

intake > demand
Thus **body mass increases**:
- may follow giving up smoking
- may reflect lack of exercise (1hour cycling or jogging to 'burn up' 100 g of chocolate)
- may have genetic basis.

intake < demand
Thus **body mass decreases**:
- may be caused by eating disorders
- may result from excessive exercise (and promotion of metabolic rate)
- may result from high thermogenesis in brown fat.

2.13 Diet and CHD

Lifestyle diseases and risk factors

Many diseases are affected by **lifestyle** (the way a person lives their life). People who pursue an unhealthy lifestyle are at greater risk of developing particular diseases than people who choose more healthy options.

Cancer and disease of the heart and blood vessels are examples of so-called **lifestyle diseases**. Today they account for the majority of deaths in the UK.

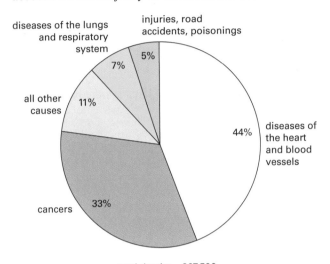

total deaths = 267 500

Causes of death in people aged under 75 in the UK

Life is uncertain. What we do might increase the chances of something happening to us. 'What we do' is our lifestyle. Choice of lifestyle can affect our health.

- If what we choose to do increases the chances of ill health, then the choice is a **risk factor**.
- Not making the choice removes the risk, so some risk factors are **avoidable**.
- **Unavoidable** risk factors are associated with events over which we have no choice. Ageing, gender, and the genes we inherit are important examples.

Diet and heart disease

People who eat too much fat and sugar may become **obese** and have a higher risk of heart disease.

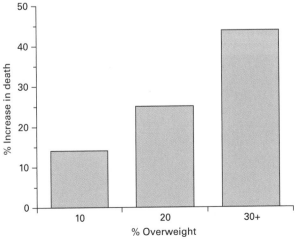

Increase in deaths from heart disease due to being overweight

Cholesterol is often associated with heart disease as it forms deposits that block blood vessels. Blocked blood vessels cause heart disease. The more cholesterol there is in the blood, the greater the risk of heart disease.

Eating food containing a lot of **saturated fat** raises the natural level of cholesterol in the blood.

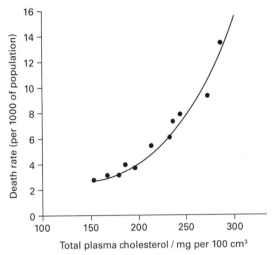

The relationship between blood cholesterol levels and risk of death from coronary heart disease

Cholesterol is insoluble and transported in the blood bound to **lipoprotein**. The proportions of **low density lipoproteins** (LDL) and **high density lipoproteins** (HDL) in the blood affect the risk of a person developing heart disease.

- Raised levels of **LDL** increase the risk of heart disease. It speeds up the formation of deposits blocking blood vessels.
- Raised levels of **HDL** lower the risk of heart disease. It removes cholesterol from the lining of the artery wall.

Reducing risk

An important first step to reducing levels of blood cholesterol is to change diet – increasing the intake of unsaturated fats at the expense of saturated fats. Different fats contain different proportions of saturated and unsaturated fatty acids.

What is CHD?

CHD stands for **coronary heart disease**, which occurs when the coronary arteries supplying the heart muscle become narrowed or blocked and can no longer supply enough oxygen for respiration. Partial blockage can cause **angina**, which is chest pain on exercising, while complete blockage leads to a heart attack or **myocardial infarction**. The heart muscle dies or becomes permanently damaged.

Questions

1 List unavoidable factors which affect the risk of someone developing coronary heart disease.

2 Explain how low density lipoproteins and high density lipoproteins affect the risk of a person developing heart disease.

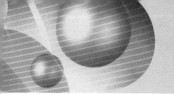
Food chains and food webs

Different words are used to describe the feeding relationships between the plants and animals of a community.

- **Producers** make food (sugars). Plants are producers. So too are some types of single celled organisms and algae. They make food by **photosynthesis**. Food chains and food webs always begin with producers.
- **Consumers** take in food (feeding) already formed. Animals are consumers.

A **food chain** shows the links between producers and consumers. It describes one pathway of food through a community. For example:

maize ⟶ chicken ⟶ human

Food chains depend on plants

Human nutrition depends on food chains, and crop plants are the basis of all our food chains.

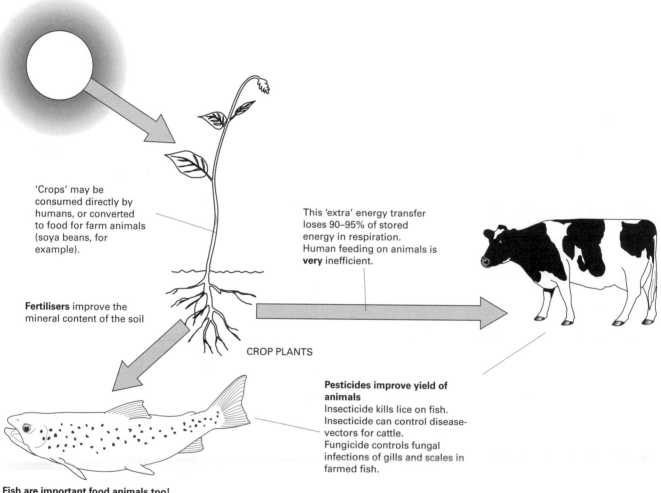

'Crops' may be consumed directly by humans, or converted to food for farm animals (soya beans, for example).

This 'extra' energy transfer loses 90–95% of stored energy in respiration. Human feeding on animals is **very** inefficient.

Fertilisers improve the mineral content of the soil

CROP PLANTS

Pesticides improve yield of animals
Insecticide kills lice on fish.
Insecticide can control disease-vectors for cattle.
Fungicide controls fungal infections of gills and scales in farmed fish.

Fish are important food animals too!
- Salmon species popular – high conversion ratio and high sale price!
- High protein food available ('trash' fish): often has artificial colourant added to it.

Selective breeding

In **selective breeding** individuals with desirable characteristics are chosen and bred together to produce new strains that have improved productivity.

Selective breeding reduces the number of different genes in the populations of crops and domesticated animals raised for food.

- The development of high yielding breeds of domesticated animals and strains of plants has reduced genetic diversity in their populations.

  ⓔ As a result the members of a population are genetically similar.

- Should the conditions in which the animals and crops are raised alter in the future, the selective breeding of new varieties able to survive the changed conditions may not be possible because there is a smaller pool of genes to draw from.

The reduction of genetic diversity and the limitations this might bring to the selection of new varieties of crops and domesticated animals are some of the ethical issues involved in selective breeding programmes.

Selective breeding produces varieties of dogs which may look very different from one another. However, the different varieties all belong to the same species and genetically are very similar.

Breeding blight-resistant potatoes

In 1845 the most popular variety of potato grown in Ireland was the Lumper. Similar in appearance and genetic make-up to the original varieties from Peru, the Lumper was partly responsible for the increase in Ireland's population during the first part of the nineteenth century. It produces excellent yields and provided a staple part of the diet for many years.

However, the Lumper has a fatal genetic flaw which no-one at the time was aware of. It is a late maincrop variety which is particularly susceptible to blight – a fungus (*Phytophthora infestans*) – which was then unknown in Ireland.

Cool wet weather in July 1845 provided the ideal conditions for the spread of potato blight spores. Blight devastated the crop of Lumper potatoes, and by 1847 the Irish potato harvest had failed completely.

The rapidly growing human population depended on this one type of food, so the blight resulted in widespread famine. More than a million people died of starvation. A further million escaped famine by emigrating, mainly to the USA. Within a few years the population of Ireland was halved.

The European potato industry was founded on varieties from tubers brought from Peru to Spain in the 1570s and to England around 1590. Since then cross-breeding has mixed up the genes contained in the original introductions and recombined them into new varieties. Breeders and growers have then selected those varieties with the most promising characteristics. However, even modern varieties are all rather similar genetically as they all rely on the limited number of original genes.

Energy and food production

Farms are ecosystems with people as consumers in a food chain of crops and livestock. The amount of food produced depends on

- energy *input*: sunlight and fuel oil
- energy *output*: efficiency of the conversion of energy input into the energy content of the food produced (crops and livestock)

The different practices which help to maximize the amount of food produced are what is called **intensive farming**.

Farming practices

Different farming practices aim to increase productivity by:

- maximizing the rate of photosynthesis and therefore the growth of crops
- reducing losses in productivity because of
  - weeds which compete with crops for space and nutrients
  - animals (mainly insects) which eat crops and spread plant disease
  - fungi which cause plant diseases
  - the dissipation of energy raising livestock

Pesticides

Pests are organisms that reduce productivity by destroying crops and harming livestock. Pesticides are substances that kill pests and therefore help to increase productivity.

- *Insecticides* kill insects.
- *Herbicides* kill weeds.
- *Fungicides* kill fungi.

Pesticides are applied to crops as sprays, fogs, or granules and to livestock as dusts or dips.

Fertilizers

The growth (and therefore productivity) of crops depends on elements and compounds that occur naturally in soil. Substances which add essential elements to soil are called **fertilizers**. They help to increase productivity by replacing the essential elements that crops take from the soil during the growing season.

The table lists the ions of some of the elements that crops need in relatively large amounts.

The ions of other elements are needed by plants in much smaller amounts (measured in tens of ppm or less). Many of them act as enzyme co-factors.

Element	ppm*	Ion	% of crop dry mass	Requirement
nitrogen	15 000	NO_3^-	3.5	synthesis of amino acids, proteins, and nucleic acids
potassium	10 000	K^+	3.4	enzyme co-factor, opening of stomata
calcium	5000	Ca^{2+}	0.7	formation of the plant cell wall
phosphorus	2000	PO_4^{3-}	0.4	synthesis of ATP and nucleic acids
magnesium	2000	Mg^{2+}	0.1	synthesis of chlorophyll
sulfur	1000	SO_4^{2-}	0.1	synthesis of some amino acids

*ppm = parts per million in solution

- **Natural fertilizers** (organic material such as manure and compost) are spread on soil. They help to maintain its structure. Fungi and bacteria decompose the material releasing nutrients which are absorbed by crops.
- **Artificial fertilizers** are added to soil as sprays or granules. Most supply nitrogen (N), phosphorus (P) and potassium (K) – the so called **NPK** fertilizers.

Rearing livestock

Livestock raised intensively are usually kept indoors. The aim is to reduce the dissipation of energy as heat from the animals so that they grow more quickly. The heat is released during cellular respiration.

Animal husbandry can improve productivity

strain of animal: selective breeding for animals with a high conversion ratio (i.e. most efficient transfer of food intake to body mass)

reduce movement/ keep warm: less energy consumption and so more efficient 'conversion'.

veterinary care antibiotics: reduce bacterial infection e.g. **vaccination:** less risk of viral infections **hormone supplements:** e.g. BST (Bovine somatotrophin) increases bulk in cattle, and oestrogen speeds up growth in poultry.

- Confinement of animals in pens or cages restricts their movement.
  - ® As a result cellular respiration during muscle contraction is reduced.
  - ® As a result the energy dissipated in exercise is reduced.
- Their environment (heating, lighting) is controlled.
  - ® As a result the temperature difference between the environment and animals' bodies is reduced.
  - ® As a result the energy dissipated in their keeping warm is reduced.
- Animals are fed low-dose antibiotics in their feed. This prevents them from becoming infected with diseases which would otherwise spread rapidly in crowded conditions.
  - ® As a result energy is not wasted in the animal's body on fighting disease.
  - ® The antibiotics cause the animal to gain weight more quickly.

Rearing livestock intensively is sometimes called **factory farming**. The animals gain weight more quickly than those allowed to roam outdoors free range. Productivity therefore increases.

Costs and welfare

The ethical issues arising from modern intensive farming include

- costs to the environment
- the welfare of livestock
- the use of pesticides which are poisonous and kill wildlife as well as pests – they may be a hazard to human health.
- the use of fertilizers which drain from land into water causing **eutrophication** – they may also be a hazard to human health.
- the use of modern farm machinery which works most efficiently in large open fields.
  - ® As a result the farming landscape is cleared of hedgerows, copses and woods.
  - ® As a result the biodiversity of ecosystems is reduced and genetic diversity lost.

Is it fair for livestock to be reared intensively so that we can enjoy eating more meat? Confining animals indoors causes them physical discomfort, boredom, and frustration. Some people argue this is cruel. Others claim that animals raised indoors must be content because they are safe from predators, sheltered, and eat well. However, we know that overeating in humans is a common sign of depression.

Fact file

In simple terms, estimates suggest that the practices of modern intensive farming increase productivity at a non-sun energy cost each year equivalent to more than 11 tonnes of oil for every person involved in the industry. The costs to the environment arise from the practices which increase productivity and which are the result of oil inputs.

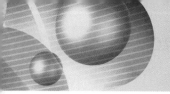

Fermentation

If a culture of yeast is supplied with glucose and water, at a temperature of around 28°C it will reproduce by **budding**.

Different strains of the yeast *Saccharomyces cerevisiae* are specialised for **baking** and for different forms of **brewing**.

Wine making from grape sugar

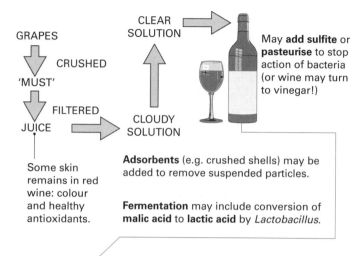

May **add sulfite** or **pasteurise** to stop action of bacteria (or wine may turn to vinegar!)

Some skin remains in red wine: colour and healthy antioxidants.

Adsorbents (e.g. crushed shells) may be added to remove suspended particles.

Fermentation may include conversion of **malic acid** to **lactic acid** by *Lactobacillus*.

N.B. Alcohol, like lactic acid, is a poison. It eventually kills the yeast cells which produce it, and does the same to human cells if taken in too large a quantity!

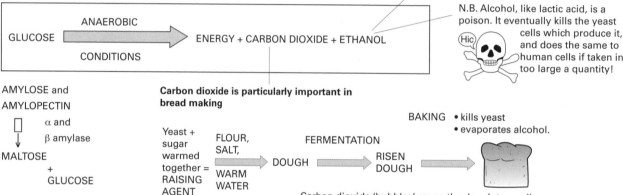

GLUCOSE — ANAEROBIC CONDITIONS → ENERGY + CARBON DIOXIDE + ETHANOL

Carbon dioxide is particularly important in bread making

AMYLOSE and AMYLOPECTIN

⬇ α and β amylase

MALTOSE + GLUCOSE

Yeast + sugar warmed together = RAISING AGENT

FLOUR, SALT, WARM WATER → DOUGH → FERMENTATION → RISEN DOUGH →

BAKING • kills yeast • evaporates alcohol.

Carbon dioxide 'bubbles' cause the dough to swell

Advantages and disadvantages of using microbes to produce food

Advantages	Disadvantages
• Microbes will often ferment waste products from other industries. • Microbes can be grown under very controlled conditions, reducing risk of contamination. • Microbes can be genetically engineered to work with great efficiency.	• There is some risk of harmful by-products. • Microbes may mutate and become less efficient.

Mycoprotein

Mycoprotein is the bodies of a filamentous fungus.

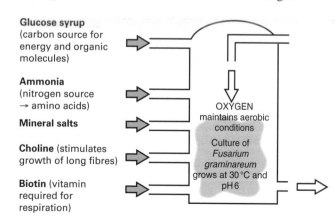

Glucose syrup (carbon source for energy and organic molecules)

Ammonia (nitrogen source → amino acids)

Mineral salts

Choline (stimulates growth of long fibres)

Biotin (vitamin required for respiration)

OXYGEN maintains aerobic conditions

Culture of *Fusarium graminareum* grows at 30 °C and pH 6

Advantages and disadvantages of mycoprotein

Advantages	Disadvantages
• Rapid production • High protein content • Low fat and salt content • High fibre content	• High RNA content can cause gout and kidney damage • Cell walls are indigestible by humans

Dairy products

Yoghurt

This is milk which has been slightly 'soured' by the excretion of lactic acid
from bacterial cultures.

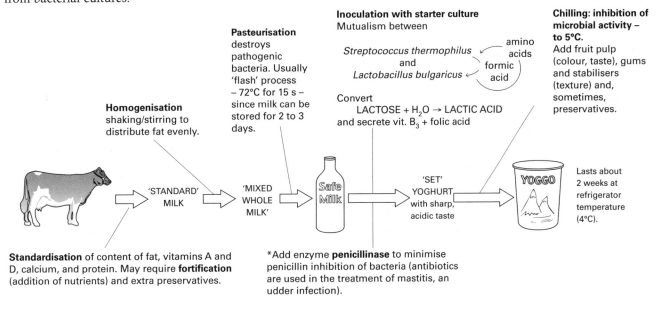

Homogenisation shaking/stirring to distribute fat evenly.

Pasteurisation destroys pathogenic bacteria. Usually 'flash' process – 72°C for 15 s – since milk can be stored for 2 to 3 days.

Inoculation with starter culture
Mutualism between

Streptococcus thermophilus and *Lactobacillus bulgaricus* ← amino acids / formic acid

Convert
LACTOSE + H_2O → LACTIC ACID
and secrete vit. B_3 + folic acid

Chilling: inhibition of microbial activity – to 5°C.
Add fruit pulp (colour, taste), gums and stabilisers (texture) and, sometimes, preservatives.

'STANDARD' MILK → 'MIXED WHOLE MILK' → Safe Milk → 'SET' YOGHURT with sharp, acidic taste → YOGGO

Lasts about 2 weeks at refrigerator temperature (4°C).

Standardisation of content of fat, vitamins A and D, calcium, and protein. May require **fortification** (addition of nutrients) and extra preservatives.

*Add enzyme **penicillinase** to minimise penicillin inhibition of bacteria (antibiotics are used in the treatment of mastitis, an udder infection).

Cheese

In its simplest form cheese is condensed yoghurt, formed by draining off
liquid whey – it is sometimes called **lactic cheese**.

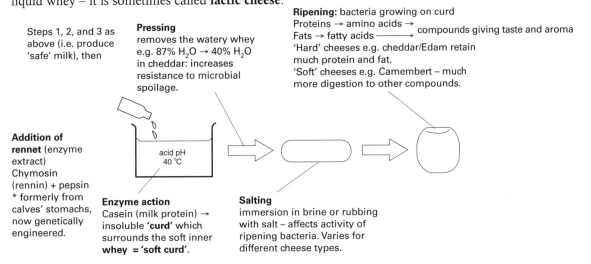

Steps 1, 2, and 3 as above (i.e. produce 'safe' milk), then

Pressing removes the watery whey e.g. 87% H_2O → 40% H_2O in cheddar: increases resistance to microbial spoilage.

Ripening: bacteria growing on curd
Proteins → amino acids →
Fats → fatty acids → compounds giving taste and aroma
'Hard' cheeses e.g. cheddar/Edam retain much protein and fat.
'Soft' cheeses e.g. Camembert – much more digestion to other compounds.

Addition of rennet (enzyme extract) Chymosin (rennin) + pepsin * formerly from calves' stomachs, now genetically engineered.

acid pH 40 °C

Enzyme action Casein (milk protein) → insoluble **'curd'** which surrounds the soft inner **whey = 'soft curd'**.

Salting immersion in brine or rubbing with salt – affects activity of ripening bacteria. Varies for different cheese types.

Why does milk go sour?

At factory milk may now contain 10^{10} bacteria /dm^3.

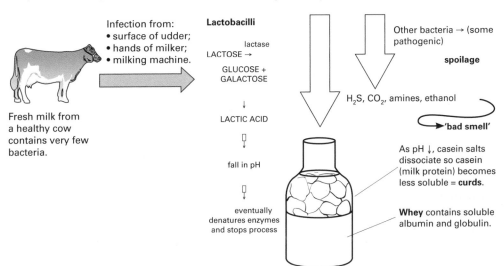

Infection from:
• surface of udder;
• hands of milker;
• milking machine.

Lactobacilli
lactase
LACTOSE →
GLUCOSE + GALACTOSE
↓
LACTIC ACID
⇓
fall in pH
⇓
eventually denatures enzymes and stops process

Other bacteria → (some pathogenic)
spoilage
H_2S, CO_2, amines, ethanol → 'bad smell'

Fresh milk from a healthy cow contains very few bacteria.

As pH ↓, casein salts dissociate so casein (milk protein) becomes less soluble = **curds**.

Whey contains soluble albumin and globulin.

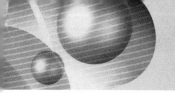

Food spoilage

Food spoilage is caused by enzymes – it can be dangerous.

Autolysis: this is the deterioration of food caused by enzymes within the food itself.

- Begins on death of animal or harvesting of crop
- Benefits: meat tenderisation and ripening of fruit
- Drawbacks: fat oxidation → rancid flavours and tastes
  'browning' of cut surfaces caused by action of polyphenol oxidase
- Natural protection by outer layers e.g. skin of apple/shell of nut.

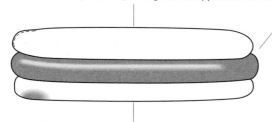

Microbial spoilage: this is caused by opportunistic organisms which can exploit a newly available food source – effects are cyclic

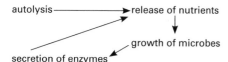

Some specific spoilage organisms

Food	Damage	Organisms
Wine, beer	souring due to ethanol oxidised to acetic acid	Acetobacter
Peanuts, cereals, fruit, bread, dried foods	aflatoxin produced causing food poisoning, mycelial growth through food	Aspergillus spp

Aflatoxin is one of the most dangerous of toxins

– haemorrhage due to leaky capillaries
– liver infection leading to hepatitis
– cancer of the liver, kidneys and colon
The fungus requires high O_2 concentration / moisture / carbohydrate source: **control by careful harvesting at high [CO_2]**

Pasteurised milk	'bitty milk' from lecithin breakdown	Bacillus cereus
Bread, fruit, vegetables	visible mycelial growth	Rhizopus

Food poisoning may result from ingested pathogens or from the release of enterotoxins onto the food source.

Pathogens then multiply within the body, causing, in particular, diarrhoea and vomiting:

e.g. Hepatitis A in shell fish
Listeria in soft cheeses and pâté
Bacillus cereus in cooked rice which is stored before use.

Enterotoxins include **Botulin** from *Clostridium botulinum*:

- One of the most lethal toxins known – 1 g might kill 100 000 humans
- Only produced under anaerobic conditions
- Paralysis, e.g. of heart due to nerve damage.

Salmonella food poisoning
(*S. typhimurium* most likely)

Intensive rearing of chickens makes it easy for *Salmonella* bacteria to spread from chicken gut to faeces to floor to chicken carcasses – then in water/on hands/on work surfaces.

68 °C : cooking to this temperature kills *Salmonella* in meat

The bacterium is not very virulent – about 10 000 000 are needed to initiate an infection. Once in the gut the organism multiplies under ideal conditions of temperature/ food availability and releases a TOXIN – this leads to a severe inflammatory response in the gut lining and fever, pain, vomiting and diarrhoea.

Treatment: antibiotics are not very effective because they do not easily reach the gut lining which is damaged.

Replace body fluids: to minimise dehydrating effects of diarrhoea.

Control by good food hygiene: washing work surfaces
cooking food thoroughly
care when handling food.

When food is spoiled

- Appearance is less attractive
- Surface dries out – reduced palatability
- Reduction in food content
- Release of enterotoxins (see above right).

N.B. Partial spoilage brings some benefits.

e.g. ripening of cheese and maturation of yoghurt.

Food preservation

Preservation methods involve inhibition of microbial activity.

Dehydration – one of the most efficient methods involves **freeze drying**: freeze rapidly then rewarm under reduced pressure → ice sublimation → **porous structure**

Benefits: better for rehydration
allied to an N_2-containing atmosphere offers 2–3 years storage.

Drawback: any fats present are easily oxidised (**rancidity**) because of open structure.

Chemical preservatives – include sulphur dioxide and sulphites, benzoic acid and benzoates and nitrates/nitrites.

- Typically act as **anti-oxidants** and inhibit auto-oxidation of fats

- Removal of O_2 gas and ↓ pH both limit growth of microbes

- SO_2/SO_2^-: widely used in sausages/dried fruits/soft drinks

- Benzoates: soft drinks

- NO_3^-/NO_2^-: meats e.g. ham/bacon are cured this way; some danger from carcinogenic **nitrosamines** ($NO_2^- + -NH_2$ in food → $-NONH_2$).

Blanching

Before vegetable foods are dehydrated (or, indeed, stored in any other way) they are blanched in boiling water or steam. This denatures enzymes such as catalase and ascorbic acid oxidase and improves both appearance and storage life of the food.

High Osmolarity

Salt or sugar can be used to generate high solute concentrations in foods. Such conditions inhibit the growth of micro-organisms which inevitably lose water by exosmosis.

Freezing

Most fresh foods contain over 60% water. If water is frozen it cannot be used by micro-organisms.
There are positive points

- very little loss of nutritive value – 'quick frozen'
- food may have more nutrients than food which is 'fresh'
- but takes 2–3 days to reach point of sale
… but some negative points too
- expansion of water → cell damage → 'mush' (e.g. strawberries)
- 'drip' from frozen food → loss of soluble nutrients (e.g. vitamin C) on thawing.

Canning – heat sterilisation kills micro-organisms and their spores. If the sterilised food is to be stored for long periods, it must be sterilised in a sealed container which will prevent recolonisation by new microbes.

Heat – steam drives out air.

Can closed – falling volume should suck ends inwards.

'Bulge' – evidence of a 'blown' can due to gas released by respiring microbes.

Cans of food must be cooled in water – any gap in the sealing might allow entry of microbes as water would be drawn inwards as the can's contents contracted. Cooling waters are almost always chlorinated to prevent accidental reintroduction of microbes.

The **botulinum cook** – the use of pressure cookers allows boiling water to exceed 100 °C. The modern HTST (high temperature, short time) technique allows a temperature of 121 °C for three minutes. These conditions will even kill the spores of *Clostridium botulinum* – this organism produces a toxin – **botulin** – which is so toxic that it is estimated that 500 g would kill almost the whole population of the UK.

Control of pH – microbes may grow more slowly under acidic conditions.

Yoghurt   Lactic acid is produced by lactose fermentation.

Pickles   Preserved by added vinegar (3% aqueous solution of ethanoic acid).

Irradiation

Sterilisation by radiation is permitted for medical supplies and drugs and also for food preservation in the UK where it is of benefit to the consumer.

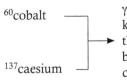

^{60}cobalt ——— γ-radiation (dose required to kill microbes is usually greater than dose permitted for humans, but should decay before consumption)
^{137}caesium ———

… but

- May be some induced radiation
- Very expensive, so use tends to be confined to high-cost foods e.g. prawns
- Some loss of vitamins A, B, C and E.

In the UK only dried herbs and spices are irradiated. Other irradiated foods may be imported. Irradiation is an effective method for reducing *Salmonella* in frozen meat.

2.17 Health and disease

What is disease?

Health is a complex subject and someone's degree of health is difficult to measure. The term 'health' is defined by the World Health Organization as

- 'a state of complete physical, mental, and social wellbeing which is more than just the absence of disease'.

Disease may be easier to visualize – we describe particular diseases with specific symptoms. Diseases may be **physical** or **mental**, and physical diseases may be **infectious** or **non-infectious**. Infectious diseases are caused by organisms that invade our bodies.

Pathogens and parasites

Organisms that cause disease are called **pathogens**. Pathogens are examples of **parasites** – they depend on another organism, the **host**, for their life functions, and cause harm to the host.

Pathogens include different types of **bacteria**, **viruses**, and **fungi**. When the human body is their host, they may make us ill. The body is warm and moist – an ideal environment for pathogens to grow, multiply, and spread.

Penetrating body surfaces

Pathogens can cause disease when they penetrate any of the body's interfaces with the environment. These include surfaces within the digestive and gas-exchange systems.

Invading pathogens must first attach to the cells of body surfaces or penetrate the cells themselves. They must also survive the host's defences against their invasion.

- Substances produced by bacteria enable them to bind with surface receptors on the cell surfaces of the host's tissues.
- Extensions of the plasma membrane enable some types of bacteria to attach to the host's tissues.
- Enzymes produced by bacteria enable the cells to enter the host's tissues. Their activity breaks down the tissues, allowing deep penetration. Other enzymes destroy the white blood cells of the host's immune system.
- A tough outer layer covers many types of bacterial cell. The layer protects the cell from the host's white blood cells. Loss of the layer makes the bacterium vulnerable to the host's defences. It is destroyed before it causes disease.

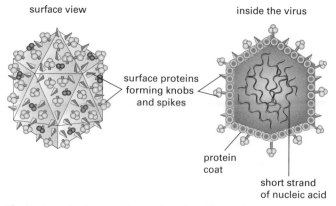

surface view

inside the virus

surface proteins forming knobs and spikes

protein coat

short strand of nucleic acid

The knobs and spikes on this virus help it to bind with the host cell.

How do pathogens cause disease?

Pathogens cause disease by producing toxins and by damaging the cells of the host's tissues. For example, bacteria produce two types of toxin – exotoxins and endotoxins.

Exotoxins are mostly proteins and very potent in small amounts. For example:

- The bacterium *Corynebacterium diphtheriae* causes diphtheria. It produces a toxin that prevents protein synthesis in the host's cells. The toxin molecule is in two parts – one part causes the toxicity; the other part promotes uptake of the molecule by host cells. The toxic part causes a swollen throat and massive internal bleeding.

Endotoxins are part of the bacterial cell surface membrane. They are less potent than exotoxins but can be lethal. For example:

- The endotoxin of *Escherichia coli* can cause a sudden decrease in blood pressure leading to septic shock.

Damage to the host's tissue may be direct or indirect.

Three major killers: malaria, HIV/AIDS, and TB

- Malaria is caused by a protoctist (single-celled eukaryotic organism) called *Plasmodium* which causes flu-like symptoms which progress to severe fever. The disease can lead to long-term weakness and often causes death. The *Plasmodium* pathogen is passed to humans by another organism called a **vector**, which in the case of malaria is the *Anopheles* mosquito. The pathogen lives in the mosquito's body and is passed to humans when an infected mosquito bites them.

- **Human immuno deficiency virus (HIV)** is the virus that causes the disease **AIDS** (autoimmune deficiency syndrome). It directly attacks the T-helper cells of the immune system, destroying them. The body's immune response to infections is weakened. People infected with HIV do not suffer from the effects of the virus itself but from the different pathogens which gain a foothold once HIV has destroyed sufficient numbers of T-helper cells.

- *Mycobacterium tuberculosis* causes **tuberculosis (TB)**. The damage to lung tissue is indirect. The bacterium triggers an immune response by the host. Phagocytic cells gather where the bacterial cells have infected cells of the lungs and release enzymes. The enzymes break down the lung cells, damaging the tissue.

> ## Questions
>
> 1 What are pathogens?
> 2 How do pathogens cause disease?
> 3 How does lung tissue become damaged when infected with the bacterium which causes tuberculosis?

Disease	Pathogen	Transmission	Global significance
Malaria	Protoctist: *Plasmodium*	Via a vector, the *Anopheles* mosquito	Widespread in tropical and subtropical regions. 225 million cases and 781 000 deaths each year: 2.23% of deaths worldwide
HIV/AIDS	Virus: human immunodeficiency virus (HIV)	By exchange of body fluids (sexual contact, infected blood, or from mother to child)	Worldwide but most widespread in Africa. Estimated 2 million deaths a year and 33 million people living with HIV.
Tuberculosis	Bacterium: *Mycobacterium tuberculosis*	By airborne bacteria in droplets of moisture from an infected person	Worldwide but majority of cases in Africa. Globally 9.4 million cases and 1.7 million deaths a year.

Transmission and global impact of malaria, HIV, and tuberculosis

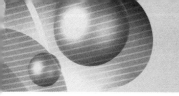

2.18 The immune response

Primary defences against disease

The skin forms a barrier to invasion of the body by pathogens. The outer epidermis is impermeable to water and to pathogens. Natural openings in the skin are protected from invasion:

- The mouth leads to the gut which is protected by hydrochloric acid in the stomach.
- Eyes are protected by lysozyme, an enzyme that destroys bacterial cell walls.
- Ears are protected by wax which kills bacteria.
- Airways are protected by cilia and mucus.
- Damage to the skin is quickly sealed by the blood clotting response.

Any pathogens that cross these barriers will enter the body and reproduce quickly in the warm moist conditions inside. The immune system of the blood then comes into play.

Responding to pathogens

When any type of 'foreign' material infects the blood or tissues, different types of white blood cell act quickly to destroy it.

- Such 'foreign' material includes viruses, bacteria, or any other cells or substances which the body does not recognize as its own.
- The white blood cells that destroy them include lymphocytes and phagocytes, and form part of the body's **immune system**.
- Their actions are the body's **immune response**.

Key words

Pathogen	Any living thing that enters the body, causing disease. Different types of bacteria and viruses are common pathogens.
Immune response	All of the reactions of the body that make an invading pathogen harmless.
Antigen	Any substance that enters the body and stimulates the development of an immune response because the body does not recognize the substance as its own. Toxins released by parasites, proteins in the plasma membrane of bacterial cells, and proteins forming the 'coat' surrounding viruses act as antigens.
Antibody	Proteins called immunoglobulins produced by the B cells of the immune system as part of the response to the presence of antigen.
Lymphocytes	Types of white blood cell. These include B cells and T cells.
B cells	These produce antibodies (**humoral immunity**) in the presence of antigens.
T cells	These *do not* produce antibodies but have a variety of effects (**cell-mediated immunity**) in the presence of antigens.
Memory cells	Lymphocytes that survive for a long period (years or even a lifetime) following an immune response. If the body is infected again by the pathogen which provoked the initial immune response, then the memory cells rapidly divide, producing new lymphocytes which destroy the pathogen before the symptoms of disease appear.
Phagocytes	Types of white blood cell which engulf bacteria, viruses, and other antigenic components, destroying them.

The response of B cells and phagocytes

B cells divide and produce clones of cells in response to the detection of antigens. Most of the cells of these clones are **plasma** cells. They produce antibodies which combine with the antigens. The combinations are called **immune complexes**.

If immune complexes form between antibodies and the antigenic proteins in the plasma membrane of bacterial cells, the bacteria clump together (**agglutination**). Agglutination facilitates (makes easy) destruction of the bacteria by phagocytes (**phagocytosis**).

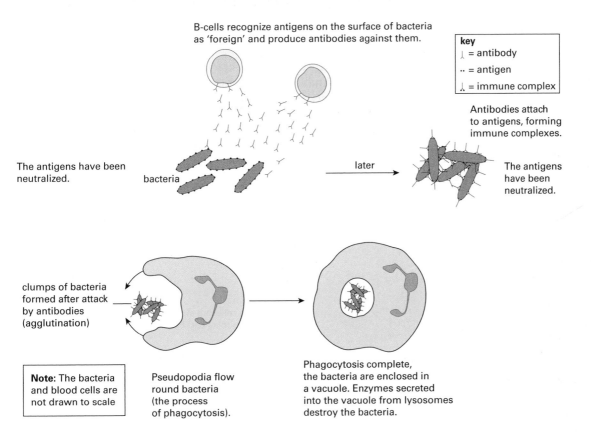

B-cells recognize antigens on the surface of bacteria as 'foreign' and produce antibodies against them.

key
⊥ = antibody
·· = antigen
⊥ = immune complex

Antibodies attach to antigens, forming immune complexes.

The antigens have been neutralized.

bacteria

later

The antigens have been neutralized.

clumps of bacteria formed after attack by antibodies (agglutination)

Note: The bacteria and blood cells are not drawn to scale

Pseudopodia flow round bacteria (the process of phagocytosis).

Phagocytosis complete, the bacteria are enclosed in a vacuole. Enzymes secreted into the vacuole from lysosomes destroy the bacteria.

Phagocytosis

Why are antibodies specific?

A molecule of antibody is Y-shaped and made up of four polypeptide chains. The sequence of amino acid units of the 'stem' of the Y is **constant** but the sequence of amino acid units of the parts that make up the 'arms' of the Y is very **variable**.

Antibodies are specific because

- millions of variations in the sequence of amino acid units of the 'arms' are possible, so millions of different shapes of antibody are possible
- when B cells recognize a particular antigen, they produce a particular antibody in response – the shape of that antibody matches the shape of the antigen. It is specific to that antigen

Antibody and antigen combine at the variable regions of the antibody (the 'arms' of the Y shape) and an immune complex forms.

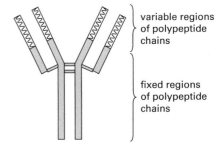

variable regions of polypeptide chains

fixed regions of polypeptide chains

The basic structure of an antibody

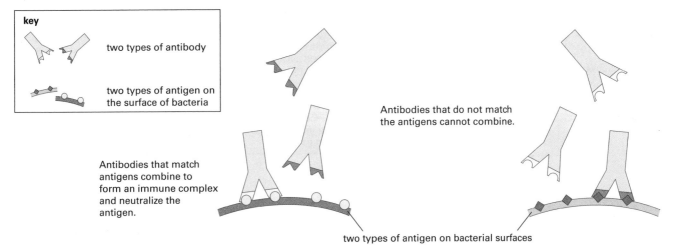

key

two types of antibody

two types of antigen on the surface of bacteria

Antibodies that match antigens combine to form an immune complex and neutralize the antigen.

Antibodies that do not match the antigens cannot combine.

two types of antigen on bacterial surfaces

An immune complex forms when the shape of the antibody and antigen match

The response of T cells

Different types of T cell have a number of roles in the immune response. For example, **T-cytotoxic** (killer) cells attach to virus-infected cells, causing them to burst (**lysis**). Antibodies prevent the virus particles released by lysis from infecting other cells. Further infection is also prevented by the virus-infected cells themselves before they burst. They release a protein called **interferon** which stops the virus from replicating.

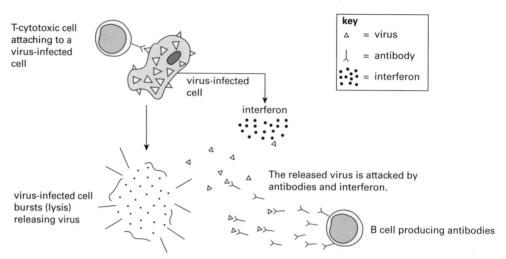

A T-cytotoxic cell attacking a virus-infected cell

Other types of T cell coordinate responses overall to the presence of antigens. The diagram summarizes the interactions between B cells, T cells, and phagocytes in the immune response.

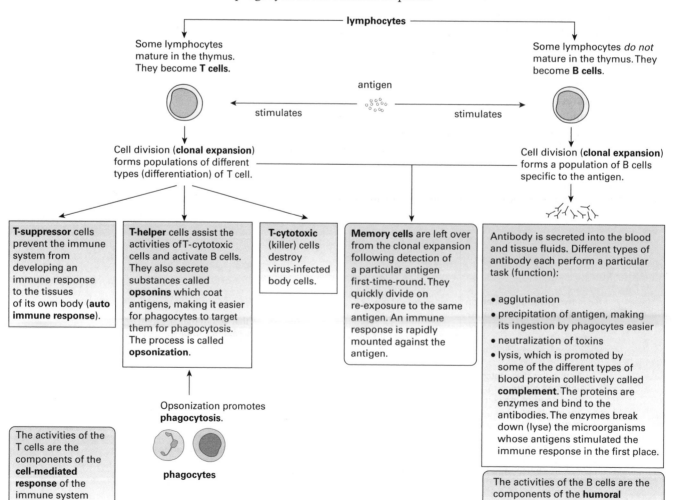

Immunological memory

On first encounter with an antigen, it takes the body a few days to produce antibodies (the **primary response**) against the infection. But on encountering the same antigens again, the response is much quicker (the **secondary response**). This is because of the presence of **memory cells**.

- B-memory cells rapidly produce plasma cells which then produce antibodies (see graph).
- T-memory cells produce plasma cells which then take part in cell-mediated immunity responses (details not needed).

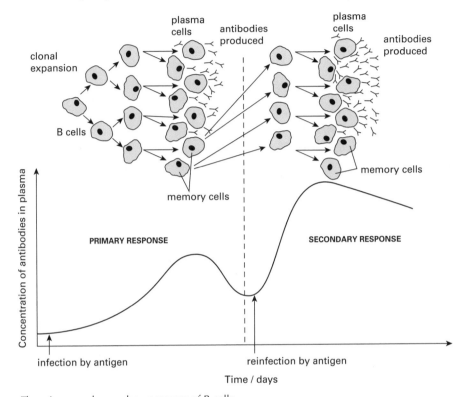

The primary and secondary responses of B cells

These enhanced responses are evidence for immunological memory, and memory cells are responsible for the effect.

Memory cells are specific for a particular antigen. This is why we do not catch mumps or chicken pox more than once in a lifetime: the rapid response as a result of immunological memory destroys the viruses which cause these diseases before they make us ill.

Antigenic variability in the influenza ('flu) virus

The surface proteins of 'flu viruses are antigens against which an infected person produces antibodies. But people may catch 'flu more than once during their lifetime. So why are the antibodies ineffective against 'flu virus when it next invades the body? After all, the production of memory cells means that we rarely catch diseases like chicken pox and measles more than once.

Unfortunately, frequent mutation means that 'flu virus antigens often change shape.

- Minor changes are called **antigenic drift**. They produce new **strains** of virus which are probably responsible for the frequent occurrence of 'flu **epidemics**.
- Major changes are called **antigenic shift** and result in new **types** of virus. They are less frequent and seem to be linked to the 10–20 year cycle of worldwide **pandemics**.

Antigenic drift and antigenic shift means that antibodies produced against a particular type of 'flu virus do not protect the person from infection by new variants of the virus. It also means that producing an effective vaccine is difficult.

Questions

1 What is an immune response?

2 Describe the different responses of B cells and T cells in the presence of an antigen.

3 Explain the difference between antigenic drift and antigenic shift in the 'flu virus.

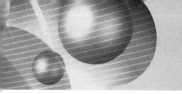

Medicines and biodiversity

Vaccines can give protection against some infectious diseases, but they are expensive and it is not practical to develop and deliver vaccinations against all diseases, especially if their effects are not usually serious for the majority of people.

Many infectious diseases are treated with **antibiotics**. These are naturally occurring chemicals used as drugs, that selectively kill pathogenic bacteria in the body. Antibiotics and other active chemicals are secreted by living organisms such as plants or bacteria, and these chemicals, or synthetic derivatives of them, are developed as drugs.

Because pathogenic microorganisms are evolving all the time, changing their genes and often mutating to become resistant to existing treatments, there is a constant and increasing need to develop new antibiotics. There are many potential new antibiotics and other drugs that have yet to be discovered in the world, and this is just one reason to maintain the biodiversity of the planet, to prevent these potentially life-saving treatments from being lost before we have discovered them.

Vaccines

Natural immunity results from the body's response to an antigen. Natural immunity may also be passed from mother to baby as antibodies in breast milk. **Artificial immunity** is brought about by vaccinations. A **vaccine** is a preparation of dead or **attenuated** (weakened) pathogens or harmless components of pathogens which stimulate an immune response in a person receiving the vaccine. It can be given orally (by mouth) or by injection.

Following vaccination, the person is protected from the effects of the active form of the pathogen should it infect the body. We say that the person is **immune** to the pathogen. The immunity is **active** because the person's immune system has been stimulated to produce memory cells. These are the basis of the person's immunity. Active immunity is long-lasting.

Passive immunity comes from the injection of antibodies produced by another animal (e.g. horse). Protection against the effects of a particular pathogen is immediate. However, protection is short term.

The table summarizes the different types of vaccines.

Vaccine made from ...	Description	Examples
dead pathogens	The pathogens are killed by heat or the addition of chemicals such as formaldehyde. They do not cause disease, but the structures of the pathogen's surface molecules are preserved. The molecules can therefore act as antigens and stimulate antibody production in the person receiving the vaccine.	whooping cough vaccine Salk vaccine against poliomyelitis
weakened live form of the pathogen	Vaccines made in this way are called **attenuated vaccines**. The attenuated (weakened) pathogen infects the person receiving the vaccine, stimulating the production of antibodies. It does not cause disease.	BCG vaccine against tuberculosis Sabin vaccine taken orally against poliomyelitis
substances made from parts of the pathogen or its toxins	Inactivation makes the toxic substances (called **toxoids**) harmless. However, their role as antigens is not affected. The production of antibodies in the person receiving the vaccine is stimulated.	tetanus and diphtheria vaccines
a particular protein or small fragment of the pathogen	The protein or fragment acts as an antigen. Vaccines made in this way are called **subunit** vaccines. They avoid some of the side effects of whole vaccines.	human papilloma virus vaccine

Live attenuated vaccines are the most popular type of vaccine. Their advantages are that

- attenuated pathogens multiply in the person receiving the vaccine. Only a low dose of vaccine is therefore needed to deliver enough antigen for an effective immune reaction to occur
- live multiplying microorganisms stimulate the production of memory cells more effectively

However, their disadvantages are that

- mutation of the attenuated pathogen may make the vaccine ineffective or, very occasionally, the pathogen may revert to the disease-causing form
- live vaccines must be stored in cool conditions

The discovery of vaccination

Edward Jenner (1749–1823) was a British country doctor who did not understand immunology as we do. But he was a good scientist who tested ideas formed from everyday experience.

He learnt from local farmers that milkmaids who caught the mild disease cowpox rarely caught the much more serious and often fatal disease smallpox. During an outbreak of smallpox in the neighbourhood, Jenner deliberately infected several of his patients with cowpox. The patients soon developed cowpox but were not affected by smallpox.

Jenner took the experiment a dangerous step further. He infected a boy who had just recovered from cowpox with pus from the spots of someone suffering from smallpox. The boy did not develop smallpox. His survival added weight to earlier ideas that giving a person a mild dose of a disease protects against more serious forms of the disease.

Jenner published his results in 1798 and the work established vaccination (immunization) as a powerful weapon in the fight against disease. At first people were suspicious and it took time for the technique to be accepted. Now smallpox has been eliminated from all parts of the world and vaccines protect millions of people from a variety of diseases.

Vaccination programmes

A vaccination programme coordinated by the World Health Organization worldwide helped to eradicate the deadly disease smallpox by 1977. As far as we know, smallpox virus exists only in secure laboratory facilities in the USA and Russia. Currently, there is debate as to whether or not these laboratory samples should be destroyed.

Other diseases such as rubella (German measles), polio, measles, and typhoid are far less common than before because of international vaccination programmes.

Mass vaccination breaks the chain of infection and makes it difficult for outbreaks of disease to occur and spread. The key to success is that most people are vaccinated – the so-called **herd effect**. However, the effect would soon disappear if the number of people vaccinated fell to levels where pathogens easily spread among unprotected individuals.

The effectiveness of mass vaccination

Whooping cough vaccine is a suspension of killed *Bordetella pertussis*. It is usually given with diphtheria and tetanus vaccines in a **triple vaccine**.

Vaccination against whooping cough started in 1957. Before then about 100 000 cases of the disease were reported in the UK each year. By 1973 more than 80% of the population had been vaccinated and the number of annual cases fell to around 2400.

However, there was a scare over the safety of the vaccine and vaccinations fell to around 30% in 1975. Epidemics of whooping cough followed: there were nearly 66 000 cases in 1982 and more than 36 000 in 1986. Methods of producing the vaccine were improved.

A publicity campaign pointing out the advantages of vaccination helped to restore public confidence. The percentage of people vaccinated increased, halting further epidemics. In 1998 there were around 1500 cases of whooping cough reported in the UK; 95% of children had been vaccinated by the age of two years. Cases fell further and by 2003 there were just 386 cases in England; however, this again rose to over 1000 in 2007.

Government responses to antigenic shift in the 'flu virus

Seasonal 'flu is a highly infectious respiratory illness spread by airborne viruses which are dispersed by coughing and sneezing. In response to the variability of the 'flu virus a seasonal 'flu vaccine is developed each year. This process is coordinated by the World Health Organization in response to information about the developing strains most likely to be prevalent that year. The 'flu jab' is used throughout the northern hemisphere and gives good protection against all strains of 'flu included in the vaccine.

In the UK the 'flu jab is offered to people who would be at risk of developing serious complications if they caught the disease. These at-risk groups include pregnant women and the elderly. Health and social care workers are also offered the vaccine to protect themselves and also help prevent the disease spreading to more vulnerable people.

Questions

1 What is a vaccine?
2 Explain the difference between active immunity and passive immunity.
3 What is the herd effect?

About 100 000 people in the UK die each year due to smoking. Smoking-related deaths are mainly due to cancers, emphysema (chronic obstructive pulmonary disease), and heart disease.

Smoking and lung disease

Research begun in the 1930s gathered data which by the 1970s established the link between smoking, lung cancer, and other diseases.

> You do not need to learn these graphs, but may need to interpret them (or similar) in the exam.

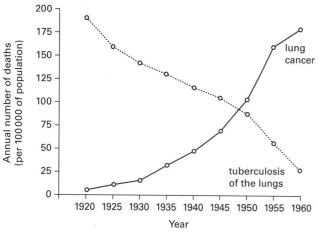

Deaths from lung disease in England and Wales from 1920 to 1960. Deaths from lung cancer increased sharply when deaths from tuberculosis fell.

- Doctors quickly realized the possible significance of the data. Many gave up smoking. Deaths from lung cancer among doctors went down compared with the population as a whole, who were less well informed.
- Further studies established the correlation between the risk of dying from lung cancer and the number of cigarettes smoked – the more cigarettes smoked, the greater the risk.

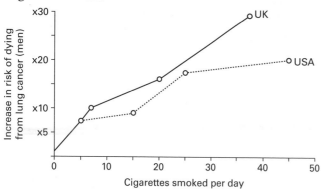

Death rates from lung cancer in men who smoke

Emphysema

Smoking is also the cause of **emphysema**, also called chronic obstructive pulmonary disease (COPD). Its symptoms include breathlessness and exhaustion.

- The substances in tobacco smoke stimulate cells (called mast cells) in the lungs to produce protein-digesting enzymes.

- The enzymes catalyse reactions which destroy the alveoli, creating enlarged cavities.
  - As a result the surface area of lung tissue available for the absorption of oxygen therefore decreases.
  - As a result oxygenation of the blood is reduced. Even a small increase in physical effort makes the person breathless and exhausted.

Smoking and cancer

Unburnt tobacco contains at least 2500 chemicals. Of these, different forms of N-nitrosamines and metal compounds have been identified as **carcinogens** (substances which cause cancer).

Different **genes** control cell division so that it stops when enough cells have been produced. **Mutations** of the genes increase their activity, stimulating cell division.

- The mutated genes are called **oncogenes**, and cell division runs out of control. Cells proliferate and a cancer develops.
- The **ras** oncogenes account for 25% of cases of lung cancer. They are activated by the carcinogens in cigarette smoke.

Tobacco carcinogens may also lead to mutations of **tumour suppressor** genes which inhibit cell division.

- Loss or inactivation of these genes contributes to the loss of control over cell division and subsequent development of cancers – especially lung cancer.
- The tumour suppressor gene called **p53** is most often involved.

Non-smokers also suffer increased risks of ill health when they breathe in smoke from other people's cigarettes – so-called **passive smoking**. The evidence supports the idea that people have the right to a smoke-free environment.

Questions

1 What are the different substances in cigarette smoke that cause cancer?

2 Why do people suffering from emphysema quickly become breathless and exhausted?

Smoking and the lungs

Bronchitis – inflammation of the linings of the bronchi

Normal

Bronchitic

- increased growth and activity of mucus-secreting cells
- loss or inactivation of ciliated epithelium.

coughing and **production of sputum**

Mechanical damage to lung tissue

+

Provision of nutrient supply for bacteria

greater risk of lung infections

- Mortality from chronic bronchitis is about 40 × higher in 25-per-day smokers than in non-smokers.

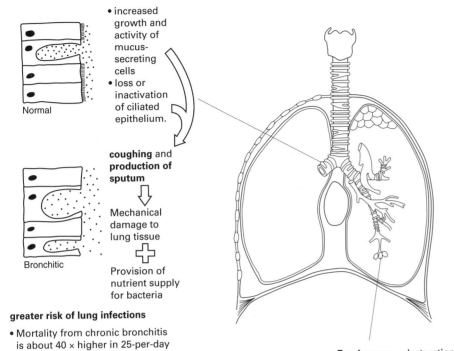

Lung cancer is very frequently caused by smoking. Common symptoms include a persistent cough, shortness of breath, coughing up blood, pain when breathing or coughing, feeling tired, and losing weight.

Lung cancer is classified according to the tissues affected:

- Small-cell lung cancer accounts for about 12% of lung cancers and is usually caused by smoking. The cancer cells are small cells and are mostly filled with the nucleus. This type of cancer can spread quite early and is likely to be treated by chemotherapy rather than surgery.

Non-small-cell lung cancer has three types:

- squamous cell cancer affects cells lining the airways, often in the centre of the lung in one of the main airways
- adenocarcinoma is cancer of the mucus-secreting cells in the airways, often in the outer areas of the lungs
- large-cell cancer – the cells are large and rounded. This type of cancer grows quite quickly.

Emphysema – destruction of lung tissue, thus reducing surface area available for oxygen uptake / CO_2 excretion:

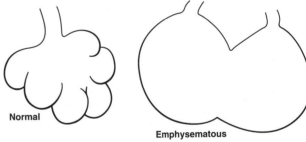

Normal

Emphysematous

- some component of tobacco smoke encourages neutrophils to accumulate in lung tissue
- neutrophils move through lung tissue by secreting enzymes (proteases and elastases) – 'repair' mechanisms, including α_1-antitrypsin, are inhibited by tobaco smoke
- infection of damaged tissue stimulates further invasion by neutrophils, making the situation worse
- emphysema is 20 × more common in 25-per-day smokers than in non-smokers.

The effects of smoking on the lungs

Smoking and heart disease

Heart disease is the biggest killer illness in the UK. About 120 000 people in the UK die each year from heart disease and about 1 in 7 of these deaths is due to smoking – in particular to inhaling **nicotine** and **carbon monoxide**.

- Nicotine raises **pulse rate** and **blood pressure**, increasing the risk to smokers of developing heart disease.
- Carbon monoxide combines with **haemoglobin** in red blood cells, reducing the capacity of the cells to absorb oxygen. The heart has to work harder to supply blood (and the oxygen it carries) to the tissues of the body.

The combined effects mean that smokers of all ages are at greater risk of dying from heart disease than non-smokers of the same age. The more cigarettes smoked, the more likely is death from heart disease.

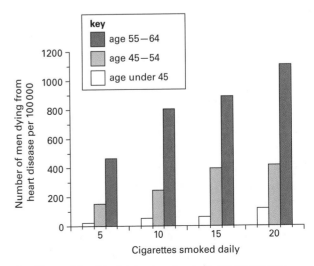

Smoking and the risk of dying from heart disease among men

93

How coronary heart disease develops

Coronary arteries

The **coronary arteries** run over and deep into the walls of the heart. They transport blood with supplies of dissolved oxygen and nutrients needed by the heart muscles.

It may seem odd that the heart needs its own blood supply, when its chambers are filled with blood. But its walls are so thick that oxygen and nutrients in the blood inside the heart would not be able to diffuse into all of the heart muscle.

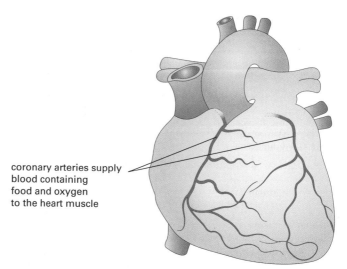

coronary arteries supply blood containing food and oxygen to the heart muscle

The coronary arteries pass over and penetrate the wall of the heart.

The smooth inner wall of healthy blood vessels allows blood to flow easily through them. Anticoagulants such as **heparin** (produced by the liver) and **prostacyclin** (produced by the lining of blood vessels) prevent blood from clotting inside vessels.

Atheroma and sclerosis of the heart

Normally streaks of fat occur in the inner layer of arteries. However, in time, fatty material (**atheroma**) may accumulate in localized deposits called **plaques**. Large amounts of **cholesterol** are found in atheroma.

As atheroma builds up it may 'harden' because of the deposition of fibrous material and/or calcium salts (**calcification**). The hardening process is called 'sclerosis' and the result is **atherosclerosis** of the arteries.

Angina

In time, plaques may narrow the arteries so much that insufficient blood reaches the tissue beyond the constriction. If the coronary arteries are narrowed, the first signs of trouble may be breathlessness and a cramp-like pain in the chest. This pain is called **angina**. It can be brought on by quick walking, anger, excitement, or anything else that makes the heart work harder than usual.

Angina is the heart's response to being starved of the oxygen that blood carries. People live with some types of angina for years, but other types get worse and may later result in a heart attack.

Heart attack

When the reduction in blood supply beyond an obstruction in the coronary arteries is so severe as to interrupt the blood supply, then the result is a **heart attack**.

- The victim may feel sick and faint, and pain usually grips the chest, spreading to the arms, neck, and jaw.
- Other signs of heart attack are sweating, breathlessness, and a pale skin because blood is not reaching the body's surface.

The events cause blockage in the blood vessels so that oxygen and nutrients cannot reach the heart muscle which the affected part of the coronary artery normally supplies. The tissue is damaged and may die.

- The clot is called a **thrombus** and the blockage a **thrombosis** – hence 'coronary thrombosis'. The term is often used to refer to a heart attack.
- **Myocardial infarction** is another term – 'myocardial' refers to the heart muscle, 'infarction' to the death of the muscle cells.

Cardiac arrest

A severe heart attack may start a rhythm disturbance called **ventricular fibrillation**, resulting in **cardiac arrest**. The electrical activity of the ventricles is so disturbed that the heart cannot pump any blood. The person becomes unconscious, and the pulse and breathing stop. It is essential to get the heart pumping again within a few minutes, otherwise the person will die.

Stroke

Coronary heart disease results from a blockage of the coronary arteries. Atherosclerosis can occur in other blood vessels too, and the blockage of an artery in the brain leads to **stroke**. A blood clot in a brain artery cuts off the supply of blood to the brain, and nearby cells die.

The possible effects of stroke include weakness of an arm or leg, twisting of the face, and problems with balance, coordination, and speech. The degree of disability is very variable depending on the area of the brain affected and the severity of the attack.

How does smoking increase the risk of heart disease and stroke?

Because smoking increases the pulse rate and the blood pressure, any blockage of the blood vessels is more likely to result in a heart attack or stroke.

The chemicals in tobacco also damage the lining of the blood vessels and affect the level of lipids in the bloodstream. This increases the risk of atheroma forming.

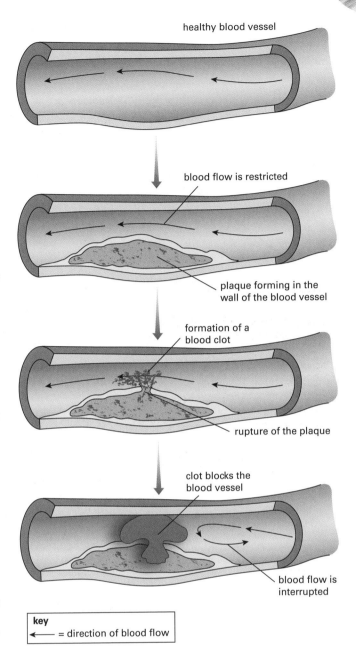

Heart attack – a plaque in the wall of a coronary artery ruptures and a clot forms.

Questions

1 The heart is filled with blood. Why does it need its own blood supply?

2 Explain the relationship between atheroma, atherosclerosis, and a heart attack.

3 Name two naturally occurring anticoagulants.

Epidemiology

Epidemiology is the study of patterns in the distribution of disease. Smoking has been closely linked to the onset of disease.

Evidence linking smoking with lung cancer

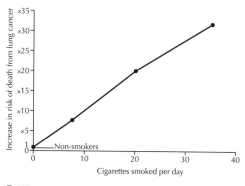

Results of a **prospective study** – looking forward and suggesting that if smoking was an important cause of lung disease and death, then early death would be more likely in smokers than in non-smokers.

A correlation between two variables does not necessarily prove that one of the variables causes the other – it might be that heavy smokers develop lung cancer because they are exposed to some other environmental factor, or that individuals with a genetic make-up that carries a risk of lung cancer are also 'genetically' more likely to smoke. The work of Richard Doll and others is so convincing because it has eliminated many other environmental factors. However, it is likely that the development of lung cancer is a complex process. It is probably multifactorial, involving both genetic and lifestyle factors.

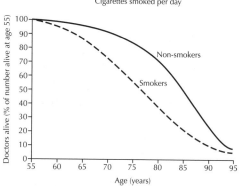

Results of a **retrospective study** – looking back at the lifestyle, occupation and environment of people who have died from lung cancer.

In 1962 the Royal College of Physicians published a report on smoking and health, which suggested a clear link between cancer and smoking. For example, among a sample of doctors living in similar-sized cities, those who smoked regularly were more likely to develop cancer of the lung.

Epidemiology has also linked smoking to low sperm counts in UK men!

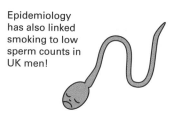

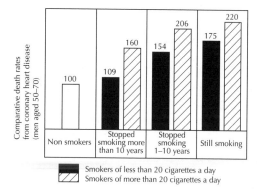

Smokers of less than 20 cigarettes a day
Smokers of more than 20 cigarettes a day

The relationship between smoking and coronary heart disease

The effect of smoking on the chance of developing bronchitis or coughing with phlegm

Epidemiology is essential! It would be unethical to force a human to smoke in order to record whether disease resulted!

2.21 Biodiversity

Uncertainties of definition

The index of diversity is not only a measure of species diversity. It is also a measure of genetic diversity. However, uncertainties of the relationship between genetic diversity and the definition of species make the use of the index of diversity as a measure of biodiversity a problem.

Sustaining the rain forests

Large-scale clearance of rain forest makes way for cattle ranches. Growing grass for the cattle to eat soon exhausts the soil of nutrients. Spreading fertilizer over such large areas of cleared forest is too expensive. The ranches are abandoned and the environment quickly turns to semi-desert.

It is possible to manage tropical rain forests for human benefit while keeping the environment intact. Perhaps the rubber-tappers of the Amazon rain forest point the way. They collect rubber, nuts, and fruits and hunt in an organized way, so that these products are not used up faster than they can be replaced naturally.

Their methods use the rain forest as a resource which renews itself in a sustainable way. In the long term, more money is made per hectare of forest than with large-scale ranching, which is unsustainable because the soil is quickly exhausted of the nutrients needed for plant growth.

Questions

1 What does the Simpson index of diversity measure?

2 How may uncertainties of the definition of species affect the reliability of the index of diversity?

3 Summarize the human impact on species diversity.

Defining species, habitat, and ecosystem

- **Species**: a group of closely related organisms capable of breeding together to produce fertile offspring.
- **Habitat**: the environment or area where an individual organism lives.
- **Ecosystem**: an environment and all the organisms living in it.

Biodiversity and the index of diversity

So far, biologists have described and named more than 5 million species of organisms. However the figure is only a fraction of the estimated millions of species yet to be discovered. So the classifications of taxonomists are inventories of the diversity of species.

The term **biodiversity** is used to describe the diversity of organisms. It can refer to a diversity of habitat, of species, or of genes within a species or within a group of species.

Species richness is the number of species present in a habitat. However, this takes no account of how many individuals of each species are present – a species with a population of 4 counts the same as a population of hundreds.

Species evenness takes account of this drawback and it compares the number of individuals of each species. It quantifies how equal the numbers are – if they are fairly even, with all species having similar numbers of individuals, then the species evenness is high; and if they are very different in number as in the example above, the species evenness is low.

One definition of species biodiversity includes the relationship between all of the species in a community of organisms and the numbers of each species. An **index of diversity** describes the relationship.

- Different species may be active within the community all of the time while others may be active only at particular times.
- The definition also includes the interactions between species.

The **Simpson species diversity index** D can be calculated from the formula

$$D = \frac{N(N-1)}{\Sigma n(n-1)}$$

Where:

D = Simpson index of diversity

Σ = sum of

N = total number of organisms of all species

n = total number of organisms of each species

The index of diversity is

- *low* in environments hostile to survival (e.g. hot deserts) where conditions may change quickly and the community is mostly affected by climate, type of soil, and other so-called abiotic factors
- *high* in environments favourable to survival (e.g. tropical rain forests) where conditions change only slowly and the community is mostly affected by interactions between species and other so-called biotic factors.

Current estimates of global biodiversity

The biodiversity of planet Earth is difficult to estimate accurately. About 5 million species are known, but the total number of species is probably much greater as there are many species that have not yet been identified.

Global biodiversity has risen and fallen in the past due to factors causing mass extinction, such as climate. Current threats to global biodiversity include natural extinction, as well as human actions such as pollution. Invasion of successful non-native species to an ecosystem can also reduce biodiversity.

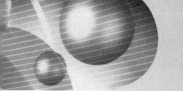

Sampling

To study the biodiversity of an ecosystem, it is necessary to estimate the population size. In other words, it will be necessary to count the number of individuals in a population. Such counting is usually carried out by taking samples (in which the organisms are in the same proportion as in the whole population) because:

1. counting the whole population would be extremely laborious and time-consuming
2. counting the whole population might cause unacceptable levels of damage to the habitat, or to the population being studied.

The samples must be representative. They should be:

1. of the same size (e.g. a 0.25 m^2 area of grassland)
2. randomly selected – for example, samples may be taken at predetermined points on an imaginary grid laid over the sampling area. The coordinates of the points may be selected using random numbers generated by a calculator
3. non-overlapping.

Quadrat sampling

Quadrats are sampling units of a known area. They are most often square, and are usually constructed of wood or metal. The quadrat can be used in simple form, or it may have wire subdivisions to produce a number of sampling points.

Reliable sampling with quadrats requires answers to three questions:

1. What size of quadrat should be used?
2. How many quadrats should be used?
3. Where should the quadrats be positioned?

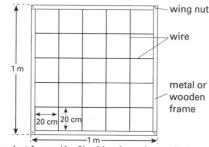

Quadrat frame (1m2) with wire sub-quadrats (each 400cm2) forming a graduated quadrat

A quadrat is used most commonly for estimating the size of plant populations, but may also be valuable for the study of populations of sessile or slow-moving animals (e.g. limpets).

A **point quadrat** is used for sampling plant populations in short grassland. It consists of pointed needles pushed through a horizontal wooden or metal frame, usually in groups of ten. Each plant touched by the point of a needle is recorded.

What size quadrat?

If individuals within a population are truly randomly dispersed, then any quadrat size should be equally efficient in the estimation of that population. However, environmental factors are rarely evenly distributed so that the living organisms dependent on them tend to occur in clumps. Small quadrats are more efficient in estimating populations (more can be taken, and they can cover a wider range of habitat than larger ones) but there are practical considerations to be taken into account (a small quadrat might not include a dominant tree in a woodland). Optimum quadrat size is determined by counting the number of different species present in quadrats of increasing size.

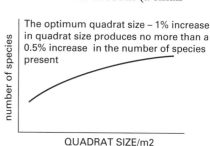

How many quadrats?

Too few might be unrepresentative, and too many might be tedious and time-consuming. To determine the optimum number, a series of quadrats of the optimum size is placed randomly across the sampling area – the cumulative number of species is recorded after each increase in quadrat number.

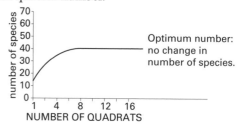

Random positioning of quadrats

The position can be chosen using random co-ordinates.

Transects

Transects are used to describe the distribution of species in a straight line across a habitat. Transects are particularly useful for describing **zonation** of species, for example around field or pond margins, or across a marsh. A simple **line transect** records all of the species which actually touch the rope or tape stretched across the habitat, a **belt transect** records all of those species present between two lines (perhaps 0.5 m^2 apart), and an **interrupted belt transect** records all of those species present in a number of quadrats placed at fixed points along a line stretched across the habitat.

A home-made quadrat frame. Two of the ten needles have been lowered.

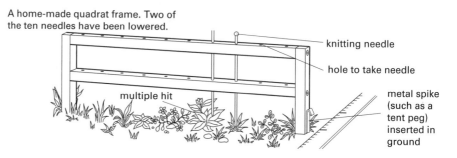

Sampling motile species

Quadrats and line transects are ideal methods for estimating populations of plants or sedentary animals. Motile animals, however, must be captured before their populations can be estimated. Once more, a representative sample of the population will be counted and the total population estimated from the sample.

Tullgren Funnel: used to collect small organisms from the air spaces of the soil or from leaf litter. The lamp is a source of heat and dehydration – organisms move to escape from it and fall through the sieve (the mesh is fine enough to retain the soil or litter). The animals slip down the smooth-sided funnel and are immobilised in the alcohol. They may then be removed for identification.

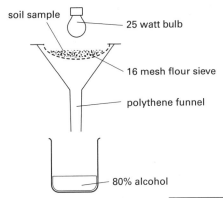

- soil sample
- 25 watt bulb
- 16 mesh flour sieve
- polythene funnel
- 80% alcohol

Baermann Funnel: works on a similar principle to the Tullgren funnel, but extracts organisms living in the soil water. The heat source drives the animals out of the muslin bag and into the surrounding water. Samples of water can be released at intervals, and the organisms in the sample collected and identified.

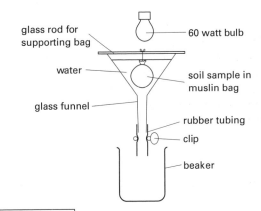

- glass rod for supporting bag
- 60 watt bulb
- water
- soil sample in muslin bag
- glass funnel
- rubber tubing
- clip
- beaker

With both Tullgren and Baermann funnels it is essential that samples are treated in identical fashion if results are to be comparative – for example, use fixed sample size, length of exposure to heat source and wattage of lamp.

Pitfall Traps: used to sample arthropods moving over the soil surface.

The roof prevents rainfall from flooding the trap, and also limits access to certain predators. The activities of trapped predators can be prevented by adding a small quantity of methanol to the trap. Bait of meat or ripe fruit can be placed in the trap.

Pitfall traps are often set up on a grid system to investigate the movements of ground animals more systematically.

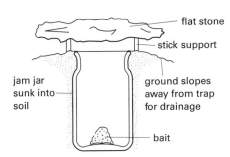

- flat stone
- stick support
- jam jar sunk into soil
- ground slopes away from trap for drainage
- bait

Pooter: used to collect specimens of insects and other arthropods which have been extracted from trees or bushes by beating the vegetation over a sheet or tray. Collection in the pooter does not harm the organism, and it can then be returned to its natural habitat.

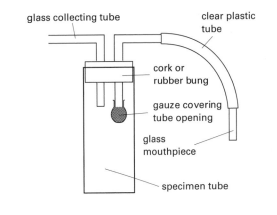

- glass collecting tube
- clear plastic tube
- cork or rubber bung
- gauze covering tube opening
- glass mouthpiece
- specimen tube

Other methods of collection are numerous. Many are based on some form of netting – for example, large mist nets may be used to collect migrating birds for identification and ringing, and sweep nets may be used to capture aerial or aquatic arthropods.

Fact file

The Swedish scientist Carl von Linné (better known as Carolus Linnaeus) devised the binomial system for naming organisms. He also laid the foundations of the system of classification we use today.

Classification

Organisms which have characteristics in common are grouped together. Placing organisms into groups is called **classification**.

Some characteristics are unique to a group – there is no overlap with other groups. Other characteristics are shared with other groups. Groups therefore combine to form larger groups forming a hierarchy of groups (or **taxa**). In the **five kingdom system**, the largest group of all is the **kingdom**.

Each

- kingdom includes a number of **phyla** (plural)
- phylum (singular) includes a number of **classes**
- class includes a number of **orders**
- order includes a number of **families**
- family includes a number of **genera** (plural)
- genus (singular) includes one or more **species**.

An alternative to this system has three **domains**:

- bacteria   • archaea   • eukarya.

Taxonomy

The term **taxonomy** refers to the strict methods and rules of classification. For example the **genus** and the **species** identify the individual living thing. Humans belong to the genus *Homo* and have the species name *sapiens*; barn owls are called *Tyto alba*.

Since the scientific (Latin) name of each species is in two parts, the method of naming is called the **binomial system** ('bi' means 'two'). Notice that

- the genus name begins with a capital letter
- the species name begins with a small letter
- the whole name is printed in *italics*

Phylogeny

The more characteristics that organisms have in common, the closer is the relationship between the individuals. By 'relationship' we mean the characteristics individuals have as a result of a shared evolutionary history which links them to a common ancestor. The term **phylogeny** refers to the evolutionary history of organisms.

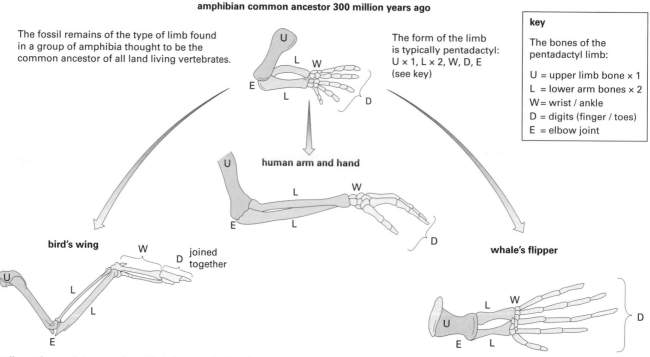

amphibian common ancestor 300 million years ago

The fossil remains of the type of limb found in a group of amphibia thought to be the common ancestor of all land living vertebrates.

The form of the limb is typically pentadactyl: U × 1, L × 2, W, D, E (see key)

key

The bones of the pentadactyl limb:

U = upper limb bone × 1
L = lower arm bones × 2
W = wrist / ankle
D = digits (finger / toes)
E = elbow joint

human arm and hand

bird's wing   joined together

whale's flipper

Different forms of the pentadactyl limb have evolved in descendants of the common ancestor of vertebrates over many millions of years.

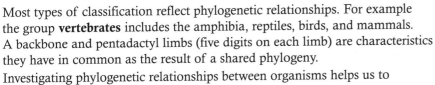

Most types of classification reflect phylogenetic relationships. For example the group **vertebrates** includes the amphibia, reptiles, birds, and mammals. A backbone and pentadactyl limbs (five digits on each limb) are characteristics they have in common as the result of a shared phylogeny.

Investigating phylogenetic relationships between organisms helps us to understand the evolution of life on Earth.

What is a species?

The reply to this question often depends on who gives the answer.

- An **ecologist** might reply that all the members of a population belong to the same species. The individuals are very similar to each other and can sexually reproduce offspring which are themselves able to reproduce.
- A **geneticist** might reply that individuals which have a very similar set of genes are members of the same species.
- A **molecular biologist** might reply that individuals which have a very similar genome are members of the same species.

In a sense, all these replies are correct – but the accuracy of their definitions depends on the meaning of 'very similar'. In the case of the ecologist's reply, for example, members of some closely related species are able to breed and reproduce fertile offspring.

Ultimately the difference between species arises from the differences between their genes. But this leads to another question:

- How great must the difference be before individuals are no longer varieties of the same species but different species?
- For example, genetic diversity between the millions of different species of insect is no more than the genetic diversity between the varieties of *Escherichia coli* – a single species of common bacteria found in the human gut!

The problem of definitions means that there is no completely satisfactory answer to the question 'What is a species?'

What's in a name?

Sometimes a living thing has several different everyday names which describe it.

- For example, the plant in this picture is called 'cuckoo pint', 'lords and ladies', 'parson-in-the-pulpit' and 'wake-robin' in different parts of the UK.

On the other hand, different living things are sometimes given the same name.

- For example, the robin in the USA is different from the robin in the UK.

Everyday names cause confusion.

- For example, British and American ornithologists (an ornithologist is someone who studies birds) would not be certain they were communicating information about the same species if they spoke to each other about 'robins'.

Questions

1 Explain the difference between the terms classification and taxonomy.
2 Why do most systems of classification reflect the phylogeny (evolutionary history) of organisms?
3 Describe the pentadactyl arrangement of your left arm and hand.

The five kingdom system of classification

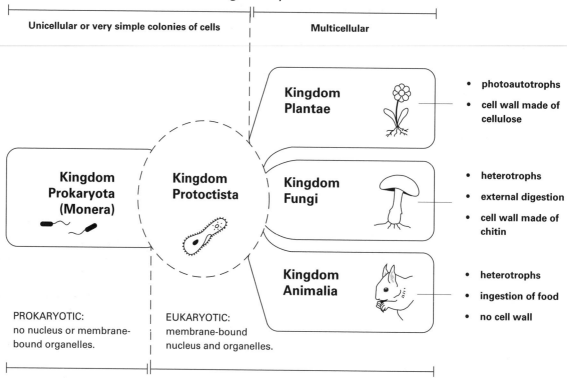

Unicellular or very simple colonies of cells | Multicellular

Kingdom Plantae
- photoautotrophs
- cell wall made of cellulose

Kingdom Prokaryota (Monera)

Kingdom Protoctista

Kingdom Fungi
- heterotrophs
- external digestion
- cell wall made of chitin

Kingdom Animalia
- heterotrophs
- ingestion of food
- no cell wall

PROKARYOTIC: no nucleus or membrane-bound organelles.

EUKARYOTIC: membrane-bound nucleus and organelles.

Sponges are an exception!

- made of many cells;
- cells may be specialised, e.g. for feeding or reproduction;
- all cells are enclosed in a form of skeleton.

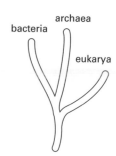

What about viruses?

- made of a protein coat surrounding RNA or DNA;
- can only reproduce inside living cells.

So could be:

- a collection of molecules, organised but not alive;
- a highly evolved parasite which originated as a prokaryote.

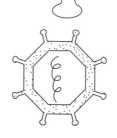

An alternative to the five kingdom system: the three domain system

- uses genetic data to divide the prokaryotes into two groups
- two groups are the **bacteria** and the **archaea** ('ancient ones')
- all other organisms are placed into one group, the **eukarya** (including animals and plants).

bacteria   archaea

eukarya

Keys and classification

A key enables identification of an organism by observation of its characteristics. Close observation allows a series of questions (the branch points in this key) to be answered, eventually leading to the organism being studied.

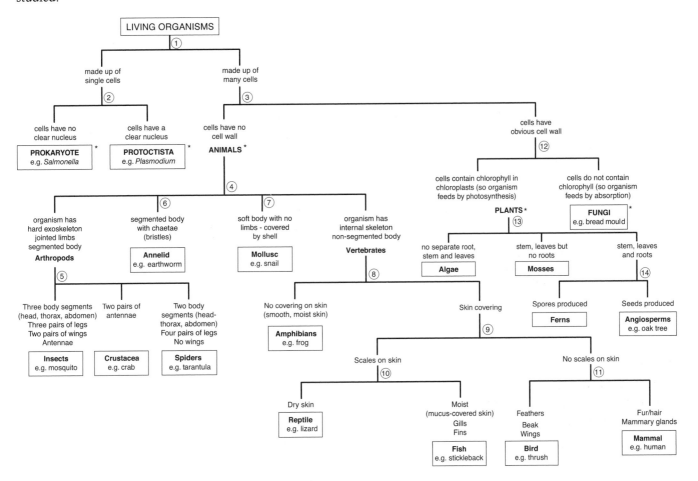

What are you? Follow the branch points at 1, 3, 4, 8, 9, and 11 to identify yourself as a mammal.

Variation

The term **variation** refers to the differences that exist between organisms. Variation occurs between different species – humans and dogs have many different characteristics, for example. Variation also occurs between individuals of the same species – for example variations in colour of eye, skin, hair, and shapes of face make human individuals different from one another.

Continuous variation

Some characteristics show variations spread over a range of measurements. Height is an example. All *intermediate* heights are possible between one extreme (shortness) and the other (tallness). We say that the characteristic shows **continuous variation**.

The distribution curve shows that height varies about a mean which is typical for the species. This is the same for any other continuously variable characteristic of a species.

Characteristics which vary continuously are usually the result of the activity of numerous sets of genes. We say that they are **polygenic** in origin.

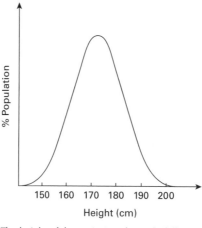

The height of the majority of people falls within the range 165–180 cm.

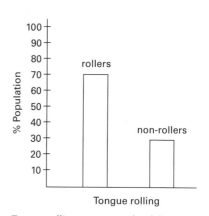

Tongue-rolling – an example of discontinuous variation

Discontinuous variation

Other characteristics do not show a spread of variation. There are no intermediate forms but distinct categories. For example most human blood groups are either A, B, AB, or O; pea plants are either tall or short (dwarf). We say that the characteristics show **discontinuous variation**.

Notice in the bar chart that some people can roll the tongue, others cannot. There are no intermediate half-rollers!

Characteristics which vary discontinuously are usually the result of the activity of one set of genes. We say that they are **monogenic** in origin.

What is standard deviation?

The distribution curve of human height has a 'middle': the apex of the graph where the height of the largest percentage of the population is identified. The value of the 'middle' is called the **mean**.

When comparing means, it is useful to know the extent to which the data is 'spread out' around each mean. The **standard deviation** is the most usual measure of 'spread-outness'.

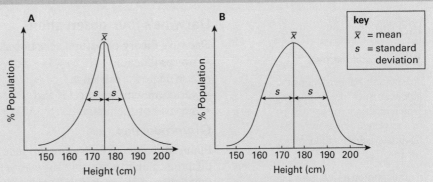

key
$\bar{x}$ = mean
s = standard deviation

Standard deviation. The mean of each distribution curve of human population is the same. However the value of the standard deviation about each mean is different. The difference shows that the variation in height of population B is greater than in population A.

Significant differences and drawing conclusions

When comparing means, it is very unlikely that they will be the same. Each mean is the average of a set of data consisting of different readings. We can never be absolutely sure of any conclusions made from comparing the means of data sets. The conclusions are **tentative** (not definitive).

However, appropriate statistical tests help us to judge if the compared means are really different from one another or that any differences are just due to chance. Most biological studies set a **confidence level** of 95% or better that the means are really different from one another. If the results meet this confidence level, then we say that the means are **significantly different**.

Environmental causes of variation

Variation arises from environmental causes. Here 'environmental' means all of the external influences affecting an organism. Examples are:

- **Nutrients** in the food we eat and minerals that plants absorb in solution through the roots. For example, in many countries children are now taller and heavier, age for age, than they were 50 or more years ago because of improved diet and standards of living.
- **Drugs** which may have a serious effect on appearance. For example, thalidomide was given to pregnant women to prevent then feeling sick and help them sleep. The drug can affect development of the fetus and some women prescribed thalidomide gave birth to seriously deformed children.
- **Temperature** affects the rate of enzyme-controlled chemical reactions. For example, warmth increases the rate of photosynthesis and therefore improves the rate of growth of plants.
- **Physical training** uses muscles more than normal, increasing their size and power. For example, weight-lifters develop bulging muscles as they train for their sport.

Variations that arise from environmental causes are not inherited because sex cells are not affected. Instead the characteristics are said to be **acquired**. Because the weight-lifter has developed bulging muscles does not mean that his/her children will have bulging muscles unless they take up weight-lifting as well!

Genetic causes of variation

Variation also results from genetic causes. Both genetic and environmental causes affect the structure and function of individuals. But only variations in features arising from genetic causes are passed on to offspring.

Genetic variation results from

- sexual reproduction – every egg and sperm is different, and each new individual arising from sexual reproduction is unique (apart from identical twins)
- mutation – spontaneous changes in the genes of a cell that can lead to new characteristics.

Questions

1 Explain the difference between a characteristic that shows continuous variation and one that shows discontinuous variation.

2 Data are often represented by their mean and standard deviation. Explain the relationship between these two measures.

3 Why are genetic causes of variation inherited and environmental causes not inherited?

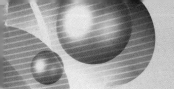

Fact file

What are adaptations?

Adaptation refers to all of the characteristics of an organism which enable it to survive in a particular environment. Characteristics include an organism's:

- **morphology** – body structure

- **molecular biology** – the organization and function of molecules in cells

- **physiology** – the way the body works

- **biochemistry** – the chemistry of cells

- **behaviour** – an individual's reactions to changes in its environment

Darwins's four observations

Darwin's theory of natural selection depends on:
- overproduction
- a struggle for existence
- variation within a species and
- survival of the fittest.

Overproduction

Plants and animals in Nature produce more offspring than can possibly survive, yet the population remains relatively constant. There must be many deaths in Nature.

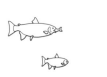

Struggle for existence

Overproduction of this type leads to **competition** – for food, shelter and breeding sites, for example. There is thus a **struggle for existence**. Those factors in the environment for which competition occurs represent **selection pressures**.

Variation

Within a population of individuals there may be considerable **variation** in genotype and thus in phenotype.

Survival of the fittest

Variation means that some individuals possess characteristics which would be advantageous in the struggle for existence (and some would be the opposite, of course).

Those possessing the best combination of characteristics would be more competitive in the struggle for existence: they would be more 'fit' to cope with the selection pressures imposed by the environment. This is **natural selection** and promotes **survival of the fittest**.

If variation is **heritable** (i.e. caused by an alteration in genotype) new generations will tend to contain a higher proportion of individuals suited to survival.

Stabilising and directional natural selection

Stabilising selection

Stabilising selection favours intermediate phenotypic classes and operates against extreme forms – there is thus a decrease in the frequency of alleles representing the extreme forms.

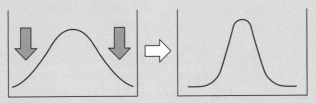

Stabilising selection operates when the phenotype corresponds with optimal environmental conditions, and competition is not severe. It is probable that this form of selection has favoured heterozygotes for **sickle cell anaemia** in an environment in which **malaria** is common, and also works against **extremes of birth weight in** humans.

Directional selection

Directional selection favours one phenotype at one extreme of the range of variation. It moves the phenotype towards a new optimum environment; then stabilising selection takes over. There is a change in the allele frequencies corresponding to the new phenotype.

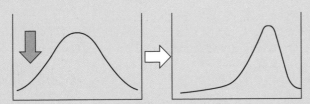

Directional selection has occurred in the case of the peppered moth, *Biston betularia*, where the dark form was favoured in the sooty suburban environments of Britain during the industrial revolution: **industrial melanism**.

Fact file

A species is:

'the lowest taxonomic group';

'a group of organisms which can interbreed and produce fertile offspring';

'a group of organisms which share the same ecological niche';

... and members of the same species have a very high proportion of their DNA in common.

Reproductive isolation and speciation

Basic process:

Speciation is the end product of evolution

Variation within a population → **Adaptation** → **New species**

Evolution by natural selection.

Reproductive isolation: prevents mixing of alleles from different populations.

Mechanisms of speciation

Geographical isolation takes place when two populations occupy two different environments which are separated by some physical barrier, such as a mountain range, a river or even a road system;
e.g. eastern and western races of the golden-mantled rosella, an Australian parakeet.

Mechanical isolation takes place when the reproductive structures are physically incompatible,
e.g. a Great Dane will not mate with a Chihuahua, and some flower species cannot be entered by the same pollinating insect.

BLEEP

FLASH

Ecological isolation takes place when two species or populations occupy different habitats within the same environment.
e.g. Marbled cat and Asiatic Golden cat may occupy the same forest, but the former is almost completely arboreal, while the latter hunts deer and rodents on the forest floor.

Behavioural isolation takes place when two different species or populations evolve courtship displays which are essential for successful mating,
e.g. Peahen is only stimulated to mate by peacock's visual courtship display; fireflies' flight patterns and flash displays prevent interspecific mating.

- **Allopatric** (= 'other country') **speciation** occurs when populations occupy different environments.
- **Sympatric** (= 'same country') **speciation** occurs when populations are reproductively isolated within the same environment.

2.26   Evidence for evolution

Fossils

- Remains of dead animals and plants which pressure, mineralisation and poor conditions for decomposition have turned to stone.
- Sedimentary rocks have been formed at the bottom of ancient seas. Oldest rocks (and fossils) are found in the deepest layers.
- Collection of fossils from different layers gives a time-limit for structural adaptations.
- Important examples are evolution of horse (limbs and teeth), development of the inner ear, links between monocot/dicot plants … and **humans**.

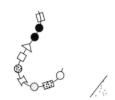

Proteins

- These are composed of long chains of amino acids.
- Some are found very widely through animal and plant groups.
- The number of amino acid changes between the same protein from different species generally shows how closely the species are related in evolutionary terms.
- Important examples are haemoglobin, cytochrome C and a hormone similar to prolactin.

DNA

- polynucleotide chain
- has 'non-coding' sections
- comparison of base sequences in non-coding regions provides a genetic clock for the time of divergence of similar species.

Other important evidence

- ATP is the universal energy currency
- phospholipid membranes are common to all organisms
- the same bases are found in the DNA and RNA of all organisms.

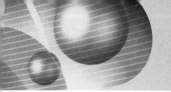

Types of antibiotic

Before the discovery of antibiotic drugs in the 1920s, bacterial diseases killed many people. Thanks to antibiotics most bacterial diseases up to now can be cured.

There are two sorts of antibiotic:

- **bactericides** like penicillin which kill bacteria
- **bacteristats** like tetracycline which prevent bacteria from multiplying

Different antibiotics affect bacteria in different ways.

- Bactericides mostly damage the structure of the bacterial cell. For example some bactericides prevent the formation of the bacterial cell wall. Osmosis floods the cells with water. The increase in hydrostatic pressure bursts the cells, resulting in **osmotic lysis**.
- Bacteristats mostly disrupt the chemical reactions taking place in the bacterial cell.

The diagram shows how different antibiotics affect bacteria.

Resistance

Genes are particular sections of DNA which control the synthesis of proteins, and so control the characteristics of organisms.

- Mutations in DNA alter genes, and so alter the proteins (and characteristics) for which the DNA codes.
- The genetic material in bacteria is DNA. Mutations in bacterial DNA may lead to new characteristics, including the development of resistance to antibiotics.

Bacterial resistance to antibiotics appeared soon after their first use to treat diseases caused by bacteria. Today some diseases, including tuberculosis, are difficult to treat with antibiotics. Bacteria resistant to all currently used antibiotics can only be treated with experimental, and possibly very poisonous, alternatives.

Unless the problems of antibiotic resistance are detected early on, then bacterial diseases which previously could be treated may become untreatable. For example, *Staphylococcus aureus* causes boils and food poisoning. Hospital strains of the bacterium are resistant to virtually all known antibiotics.

How does resistance develop?

Some types of bacteria are inherently resistant to antibiotics. For example, the capsule which surrounds some types of bacterial cell is a barrier to antibiotics.

Other types of bacteria, previously sensitive to antibiotics, may acquire resistance through changes in the bacterial genetic material. The process is driven by two processes:

Mutation

- Mutated genes which confer resistance are inherited by offspring from parent cells in which the mutation occurred.
- The mutation may spread very rapidly because a new generation of bacterial cells inheriting the mutation may be reproduced every 20 minutes or so.
- **Vertical gene transmission** refers to the inheritance of resistance genes from parental cells to offspring.

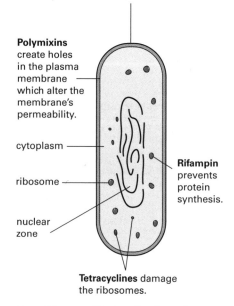

Penicillin and **cephalosporins** prevent the bacterium from making components of the cell wall, which is therefore weakened. The bacterium is then more easily destroyed by immune reactions.

Polymixins create holes in the plasma membrane which alter the membrane's permeability.

cytoplasm

ribosome

nuclear zone

Rifampin prevents protein synthesis.

Tetracyclines damage the ribosomes.

How different antibiotics affect bacteria

Fact file

Antibiotic resistance is not the only example of evolution by natural selection that causes problems for humans. Pesticides such as insecticides, herbicides, fungicides, and rodenticides are used in agriculture to improve yields by reducing competition with crops.

Continued exposure to these chemicals provides a selection pressure. Directional selection produces strains of pests that are resistant to pesticides, and this has huge environmental and economic implications.

Fact file

Darwinian evolution through natural selection is one route by which bacterial resistance to antibiotics develops. In hospitals the selection pressure of antibiotics is so intense that strains of bacteria quickly evolve resistance to a range of antibiotics.

Exchange of resistance genes

Resistance genes are exchanged between different strains of a species of bacterium or between different species of bacteria. The genes exchanged confer resistance on the bacteria acquiring them.

- **Transduction** occurs when viruses which only infect bacteria transfer genes between two closely related bacteria.
- **Transformation** occurs when bacteria take up genes from their immediate environment.
- **Conjugation** occurs when two bacteria make direct cell-to-cell contact. Plasmids containing resistance genes may pass from one cell to the other.

Horizontal gene transmission refers to the transfer of genes between unrelated bacterial cells. The diagram summarizes the processes.

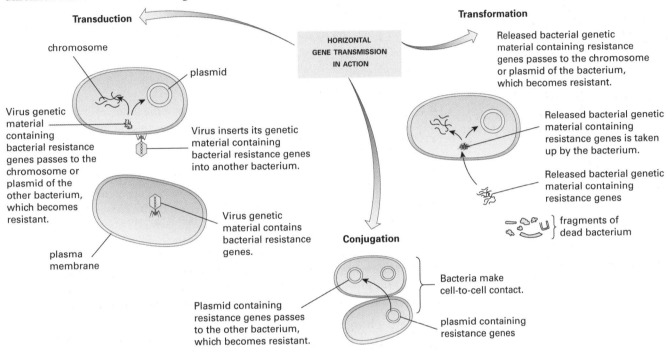

Horizontal gene transmission

How do bacteria resist antibiotics?

Different mechanisms confer bacteria with antibiotic resistance.

- Inactivation by bacterial enzymes which make antibiotics ineffective is the most common.
- Alternatively the antibiotic's target in the bacterial cell alters so that the cell is no longer affected by the antibiotic.

The diagram summarizes other mechanisms of antibiotic resistance.

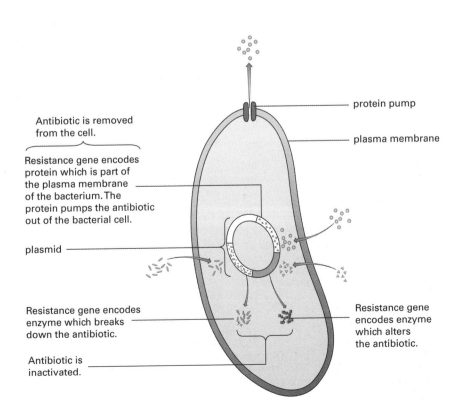

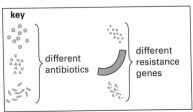

Bacterial resistance mechanisms

Background to conservation measures

Our well-being depends on keeping a balance between using resources and protecting the environments from where the resources come. Conservation enables us to

- use renewable resources (plants, animals) in a sustainable way
- reduce our use of non-renewable resources (metals, fossil fuels) through recycling and the discovery of alternative materials for the production of goods
- use land so that conflicting interests between human needs and the impact of these needs on the survival of plants and animals and their environments are reduced
- reduce pollution by the development of more efficient industrial processes, which produce less waste and use less energy
- introduce more environmentally friendly methods of farming.

Reasons for conservation

It is in the interests of the human race to preserve the biodiversity of the Earth – both plant and animal species – for many reasons, including:

- Economic reasons: biological resources are useful to us as they provide us with food, drugs, and products such as timber, dyes, and oils.
- Ecological reasons: the organisms in an ecosystem interact with each other in complex ways and if one species is lost this can upset the natural balance and have unforeseen consequences for the rest of the ecosystem.
- Ethical reasons: humans are a dominant and powerful species. We have a duty not to knowingly destroy habitats or species but to conserve them for future generations.
- Aesthetic reasons: the beauty and variety of the many diverse ecosystems on Earth provide pleasure for many people as seen by the importance of travel and tourism in undeveloped regions.

Climate change and its effect on biodiversity

Origins of greenhouse gases

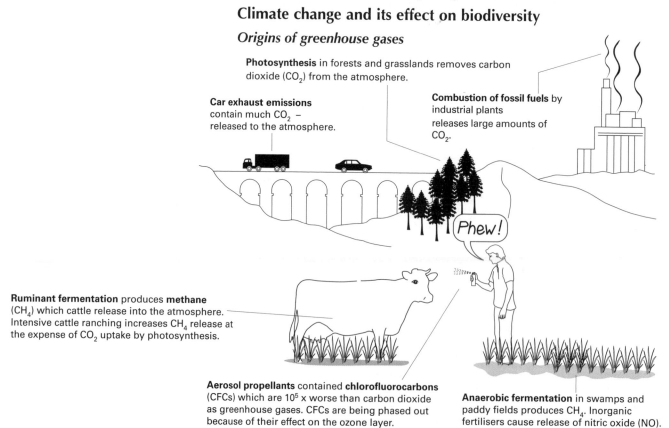

Photosynthesis in forests and grasslands removes carbon dioxide (CO_2) from the atmosphere.

Car exhaust emissions contain much CO_2 – released to the atmosphere.

Combustion of fossil fuels by industrial plants releases large amounts of CO_2.

Phew!

Ruminant fermentation produces **methane** (CH_4) which cattle release into the atmosphere. Intensive cattle ranching increases CH_4 release at the expense of CO_2 uptake by photosynthesis.

Aerosol propellants contained **chlorofluorocarbons** (CFCs) which are 10^5 x worse than carbon dioxide as greenhouse gases. CFCs are being phased out because of their effect on the ozone layer.

Anaerobic fermentation in swamps and paddy fields produces CH_4. Inorganic fertilisers cause release of nitric oxide (NO).

All living organisms release carbon dioxide by respiration – the additional greenhouse gases contributed by humans (anthropogenic contributions) include methane in addition to greater quantities of carbon dioxide.

The greenhouse effect

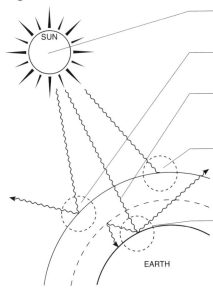

The Sun, at a temperature of 6000 °C, emits radiation which is mostly in the visible band.

About 10% of the solar energy is reflected back to space by the Earth's atmosphere.

About 83% of the solar energy penetrates the atmosphere, warms the Earth's surface and is re-emitted in the infrared range.

About 7% of short wavelength radiation helps to generate ozone.

Some of the Earth's infrared emissions are re-reflected back to the Earth's surface → **warming,** particularly by H_2O (absorbs and re-emits radiation of 4–7 μm) and CO_2 (absorbs/re-emits at 13–19 μm), **but most escapes back to space through a 7–13 μm 'window'.**

The greenhouse gases close this window and thus allow the Earth's own infrared radiation to warm its surface.

How global warming could reduce biodiversity

- **direct habitat loss e.g.**
  - melting ice reduces polar bear habitat
  - deserts (limited food chains) extend
  - flooding of many species-rich wetlands.

- **changing agricultural practices e.g.**
  - more extensive planting of monoculture cereal crops and softwood plantations
  - need to utilise 'wild' land for agriculture as drier climate → lower yields of staple crops.

- **disease vectors may extend range**
  - mosquitoes may move to the U.K.

- **pathogen life cycles may change**
  - warmer winters in Canada have allowed pine beetles to survive and kill much of northern forest.

Conservation and agriculture

Selective breeding of both plant and animal species over thousands of years means that we now rely on a relatively small number of well adapted species to provide our food. This breeding process has greatly reduced the **genome** – the total number of genes – in domestic populations. These populations are therefore less able than wild populations to adapt and withstand environmental change.

Conserving wild species maintains the pool of genes available to us for future breeding programmes to develop breeds adapted to changed environments; for example, individual agricultural areas through the world may become hotter, cooler, dryer, or wetter as a result of climate change.

As the human population grows and the demand for food increases, wild species also give us the opportunity to develop breeds suited to farming in new environments that were previously wild.

Climate change is resulting in changes in distribution of diseases, affecting not only humans but also crop plants and domestic animals. Global warming and increased rainfall affect the abundance and distribution of vectors such as mosquitoes and ticks that carry animal diseases. Increased temperatures allow mosquitoes to survive winters in areas where they would otherwise have died, leaving more mosquitoes to breed and transmit disease during the summer season. Other factors such as deforestation and natural disasters increase the incidence of vectors and hence the spread of vector-borne diseases. Fungal plant diseases also thrive in warmer wetter conditions, reducing harvests.

Wild animal and plant species have evolved along with these diseases so we can incorporate the genes from disease-resistant wild populations to provide disease resistance in new selective breeding or genetic modification programmes.

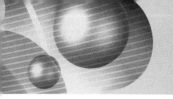

2.29 In situ and ex situ conservation methods

In situ methods: forestry

Forestry can involve in situ conservation of wildlife. 'In situ' means 'on site' conservation and is the process of conserving an endangered species in its **natural habitat**.

The habitat must be managed

- ensure food/nesting sites for endangered species
- may need to control predators or competitors (e.g. grey squirrel)
- should have the same population density as in equivalent 'natural' habitats.

'In situ' is better: endangered population has the genotype well-suited to this habitat!

Population must be large enough to allow enough genetic diversity so that endangered species can adapt and evolve over time.

DECIDUOUS, BROADLEAVED SPECIES:

- planted along edges of commercial woodlands improve the amenity value/appearance of the woodland;
- support very many more species of insect, and thus contribute more food to insectivorous species;
- native species usually fruit more easily providing an additional food source;
- can be planted to provide corridors for movement between suitable habitats for species with specific feeding or nesting requirements.

STANDS OF DIFFERENT AGES: at different stages of development trees may provide cover, food or breeding sites (rarely all three at the same time).
Thus a variety of ages of tree has optimum value for wildlife. The maximum benefit is derived from both mixed age and mixed species stands.

STACKS OF 'BRASH': the cuttings from commercial timber, or from roots extracted during forestry operations provide:

- corridors for wildlife;
- habitats for many insects;
- calling/nesting sites for song birds;
- windbreaks for the re-establishment of planted trees;
- lying-up sites for nocturnal mammals.

OPEN SPACE: clearings in woodland offer:

- 'lighter' habitat which encourages growth of an herbaceous layer;
- higher temperatures which may encourage basking reptiles and flying butterflies;
- an 'edge' effect – the greatest number of species is found along the edges of woodland.

LEAF LITTER: the fallen leaves and fruits from broad-leaved species decompose more quickly and produce a less acidic environment than the litter of 'needles' from coniferous species. This permits the germination and growth of a more biodiverse community of herbaceous plants and fungi.

DEADWOOD: fallen trees, or even some deliberately felled, are left to rot naturally.

- This provides a habitat for fungi, mosses and ferns as well as for insects and their larvae which may be significant food sources for insectivorous birds and mammals.
- Nutrients are naturally returned to the soil.

Ex situ methods: zoos and botanic gardens

Ex-situ ('off site') conservation

- involves removing part of a population from a habitat where it is threatened to a new location where it can be protected

colony relocation

(e.g. some bat species to man-made 'caves'): very difficult to recreate original environment (e.g. climate, other species present).

human care

zoos

- captive breeding
- education of public
- revenue generation (10% of world population visits a zoo each year!)
- academic research
- keep stud books and gene banks.

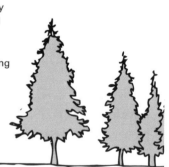

- seed banks may keep viable seeds for 100+ years
- plants may be induced to breed and produce seed by controlling environmental conditions.

Problems with ex-situ methods:

- gene pool may be limiting, reducing adaptability of species
- cannot save wild species if habitat is lost
- can be extremely expensive …
  … but it may be necessary in extreme circumstances (e.g. flooding a valley as a reservoir).

Conservation of species

There are several possible strategies, which all involve management of a habitat.

- **Preservation:** involves keeping some part of the environment **without any change**. Only possible if the area can be fenced off/protected.
- **Reclamation:** involves the restoration of **damaged** habitats. Often applies to the recovery of former industrial sites, such as mineworkings.
- **Creation:** involves the production of **new** habitats. Only really possible with small areas e.g. digging a garden pond, or planting of a new forest.

Pressures on a habitat

- Humans have a significant **biotic** impact on the environment

| **Temporary** when humans were **nomadic**: environment has periods of recovery | **Permanent** as humans became **cultivators** and **settlers** |

Use of tools and domestication of animals → more efficient agriculture/support for larger population.

Greater demands for shelter, agricultural land and fuel → greater rate of deforestation.

Development of fossil fuels → more use of machines/greater 'cropping'/larger populations.

Pollution as greater use of fossil fuels/pesticides/ fertilisers and development of nuclear power.

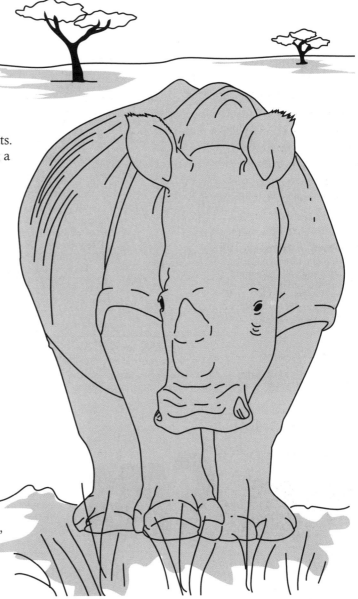

Direct protection may be necessary

Even when a suitable **habitat** (able to provide **food**, **shelter** and **breeding sites**) is available, individual species may be at direct risk from humans.

- Rhinoceros may be hunted for their horns, mistakenly believed to have medicinal or aphrodisiac properties.
- Elephant may be hunted for their ivory tusks.
- Primates (e.g. chimpanzees) and other species may be hunted as 'bush meat'.
- Butterflies, molluscs (shells) and plants may be 'collected'.
- Parrots, primates and fish may be collected for the pet trade.

Management is a compromise!

- Maintenance of a particular habitat (e.g. chalk hillside for wildflowers/butterflies) often means **halting succession**.
- The requirements for wildlife must be balanced by human demands for **resources** (e.g. mining for uranium), **recreation** (e.g. diving around coral reefs) and **agricultural land**.

Flagship species

Large, attractive and 'cuddly' species attract funding from agencies and donations from the public protection for less attractive species (e.g. beetles and worms) and live in the same habitat.

HELP!

2.30 International and local co-operation in species conservation

Conservation at many levels is needed to maintain biodiversity

The Royal Society for the Protection of Birds is primarily concerned with national issues (e.g. preservation of Caledonian pine forest in Scotland) but plays a large part in international conservation by:

- monitoring trade in birds;
- advising on the management of overseas nature reserves.

RSPB

The RSPB was founded in Didsbury, Manchester by a group of women trying to protect the Great Crested Grebe from hunters collecting feathers to decorate women's hats!

COUNCIL FOR THE PROTECTION OF RURAL ENGLAND (C.P.R.E.) is a national body which has particular concerns for local issues e.g. **the planning of roads through sensitive areas** such as ancient woodlands. They take advice from scientists who conduct **environmental impact assessments.**

GREENPEACE runs a series of non-violent, direct action campaigns such as **Protection of Marine Mammals** and the **Regulation of the Disposal of Toxic Waste.**

| International | ➕ | National | ➕ | Local |

CONVENTION ON INTERNATIONAL TRADE IN ENDANGERED SPECIES (C.I.T.E.S.) monitors trade in wildlife and wildlife products and tries to impose controls where necessary.

A licence is necessary for the import of any wildlife product.

INTERNATIONAL UNION FOR THE CONSERVATION OF NATURE AND NATURAL RESOURCES (I.U.C.N.) tries to co-ordinate action and monitor success on species survival. The I.U.C.N. publishes the Red Data Books which list in detail the animal and plant species threatened with extinction.

It identifies which species are critically endangered (CR), endangered (EN), and vulnerable (VU), so that conservation work can be channelled in the right direction.

WORLDWIDE FUND FOR NATURE (WWF) attempts to maintain **habitats** and thus the biodiversity they contain. May utilise individual **flagship species** to draw attention to and raise funding for the need to conserve habitats. These species are usually attractive mammals like the panda shown in the WWF logo.

WWF

Rio convention on biodiversity

The United Nations Conference on Environment and Development (UNCED), also known as the Rio Summit or Earth Summit, was a major UN conference held in Rio de Janeiro in 1992. It addressed many global environmental issues, including toxic wastes, climate change, and alternative energy sources. Importantly for biodiversity, it resulted in the Convention on Biological Diversity, an international legally binding treaty. Countries that sign up to the treaty agree to three main goals:

- conservation of biodiversity
- sustainable use of resources
- sharing benefits arising from genetic resources.

The work continues: 2010 was the International Year of Biodiversity. The Nagoya Protocol was adopted at a convention in Nagoya, Japan – it promotes the fair sharing of benefits arising from genetic resources.

Conservation of species

OUTSTANDING SCIENTIFIC AND POLITICAL CONTRIBUTIONS:

Diane Fossey: Gorillas

Jane Goodall: Chimpanzees

Gerald Durrell: Captive breeding of Golden Lion Tamarin

John Aspinall: Tigers

INDIVIDUALS

Local authority planning

Biodiversity Action Plan (BAP)

The Rio Summit asked individual countries to develop national strategies to identify, conserve, and protect existing biological diversity, and enhance it where possible. The UK was the first country to publish a national Biodiversity Action Plan (BAP) in 1994.

The UK BAP identifies priority species and habitats which are the most threatened and require conservation action. Species and habitats were considered against a set of criteria based on international importance, rapid decline, and high risk. Marine biodiversity, terrestrial/freshwater species, and terrestrial/freshwater habitats were considered separately in developing a revised list of 1150 species and 65 habitats in 2007. The list details the most important types of action in order to conserve each species.

Local planning authorities prepare development plans which set the framework for acceptable development in their area. They are also responsible for assessing most applications for planning permission. The BAP list is used by local authorities throughout the UK, which are required to consider biodiversity in local planning policy and planning decisions. Species, habitats, and sites are given varying degrees of protection relating to their conservation status, rarity, and wider ecological value. As well as the UK BAP list, endangered or at-risk species are also listed as Species and Habitats of Principal Importance in England under the NERC (National Environment Research Council) Act 2006. In any local area there is also a network of local wildlife sites, reserves, and other features identified as being of value to the area.

Environmental Impact Assessment (EIA)

An Environmental Impact Assessment forms part of the planning permission application of proposed major developments. An EIA gathers information to enable the local planning authority to understand the environmental effects of a development before deciding whether or not it should go ahead. EIAs should include information about the impact of the proposed development on biodiversity. An EIA is needed if the development might affect a population, a whole species, or an ecosystem of scientific, ecological, or cultural value.

Index

penicillins 81, 110
penicillinase 81
pentadactyl limb 100
peptide bond 54
peptides 54–58
pericycle 45
pesticides 76, 78–9, 110, 115
phagocytes 86–90
phagocytosis 17, 86–90
phloem 24, 26, 45–7, 51
phosphodiester bond 64
phospholipid 12–13, 63
  bilayer 12–13
phylogeny 100
phylum 100
physiology 100
Phytophthora infestans 78
pickling 83
pinocytosis 17
pitfall traps 99
placenta 43
plant cell 11
  and osmosis 19
Plantae 102
plants
  adaptations to dry conditions 50
  transport in 44–51
plaques 94–5
plasma membrane 8, 12–13, 14,
  63
plasmid 10, 11, 111
plasmodesmata 11
Plasmodium 85
plasmolysis 19
pleural cavity 31
pleural fluid 53
pleural membranes 29, 31
poisons 72
polygenic characteristics 104
polymixins 110
polypeptides 54–58, 68
polysaccharides 58–63
pooter 99
population, estimating 98–9
potassium ions, transport of 16–17
potatoes 77
potometer 48
preservation, of species 115
preservatives 83
pressure potential 19
primary immune response 89
primary structure, protein 55
producers 76
product 69–70
prokaryotes 10–11, 102
prophase 20–1
proplastid 11
prospective study 96
prostacyclin 94
prosthetic group 42
protein
  in diet 73
  and evolutionary relationships
    109
  functions 56–7
  in plasma membrane 12–13
  quality 73
  structure 54–58
  synthesis 9, 66
  testing 57
  see also carrier protein, fibrous
    proteins, globular proteins
Protoctista 102
protoplasm 8
pulmonary artery/vein 34–35, 36
pulmonary system 34–5
pulmonary ventilation 33

pulse 39
pulse rate 93
Purkyne tissue 36–7

quadrat 98–9
quaternary structure, protein 55–6

rain forests 97
ras oncogene 92
rate of enzyme–catalysed reaction
  70–31
raw data 71
receptors, membrane 13, 14
reclamation, of habitats 115
renal artery 35
renal vein 35
rennet 81
replication 67
reproduction, asexual 20, 22
reproduction, sexual 23, 105
reproductive isolation 107
residual volume 32
resistance, to antibiotics 90,
  110–11
resolution/resolving power 6–7, 8
respiration 8, 10, 16, 28, 43, 47,
  49, 61, 76, 79, 80
retrospective study 96
Rhizopus 82
ribonucleic acid *see* RNA
ribosomal RNA 65
ribosome 8, 9, 10, 11, 66
rifampicin 110
Rio Summit 116–17
risk factor 75
RNA 11, 64–6
root 28, 44, 45, 51
  hair cells 25
  hairs 28, 45
  pressure 47
roughage 73
RSPB 116

Saccharomyces cerevisiae 80
Salmonella 82–3
salting 83
sampling 98–9
saturated fat 73, 75
saturated/unsaturated fatty acids
  62
scanning electron microscope
  (SEM) 6–7
second messenger 14
secondary immune response 89
secondary structure, protein 55
selective breeding 77
semi–conservative replication 67
sexual reproduction 23
shellfish 82
sieve plate 44
sieve tube 44, 51
significant difference 105
Simpson species diversity index
  97
sink 44, 51
sinoatrial node (SAN) 36–7
skin 86
smallpox 91
smoking 92–6
sodium ions, transport of 16–17
sodium–potassium pump 16–17
solute potential 19
source 44, 51
specialised cells 24–6
speciation 106–8

species 97, 98–9, 100–1
  conservation 112–17
  evenness 97
  richness 97
specific heat capacity 52
specificity 69
sperm 24
spindle 21
spirometer 32–3
spleen 41
sponges 102
squamous cell cancer 93
squamous epithelium 25
stabilising selection 107
staining 6–7, 21
standard deviation 105
starch 59–1, 73
stem 44, 51
stem cells 24
steroid hormones 73
stomata 46, 49–50
stroke 95
struggle for existence 106
suberin 11
substrate 69–70
succession 115
sucrose 51, 58–9, 73
sugars 44, 58–9, 73, 83
surface area to volume ratio 27
surface area, and diffusion 15
surface tension 52
survival of the fittest 106
sympatric speciation 108
symplast pathway/symplastic
  route 11, 45–6
synovial fluid 53
systemic system 34–5
systole 37–38

T cells 86–90
target organ 14
taxa 100
taxonomy 100–1
T–cytotoxic cells 88
telophase 20–1
tentative conclusions 105
tertiary structure, protein 55–6
tetracyclines 110
thermoregulation 53
thoracic cavity 31
thoracic duct 41
thrombosis 95
thrombus 95
thymine, T 64–5, 68
thymus 41, 88
tidal volume 32–3
tissue fluid 39–40
tissues 25
tonoplast 11
tonsils 41
toxins 82–3, 85
toxoid 90
trachea 29–31
transcription 9, 66
transduction 111
transect 98–9
transfer RNA (tRNA) 65–6
transformation 111
translation 9, 66
translocation 44, 51
transmission electron microscope
  (TEM) 6–7, 8, 10–11
transpiration 45–48, 49
  pull 46–7
  rate 47–48
  stream 53
transport 34–5, 44–51

triglycerides 62
triple vaccine 91
tuberculosis (TB) 85
Tullgren funnel 99
tumour suppressor genes 92
turgor 50
turgor pressure 19

United Nations Conference
  on Environment and
  Development (UNCED) 116
uracil, U 65

vaccination 90–1
vaccines 90–1
vacuolar route 45–6
vacuole 11
valves
  heart 35, 36–7
  lymph vessels 40
  vein 39
variation 104–6
  environmental 105
  genetic 105, 106
vascular bundle 44
veins 34, 39
vena cava 38–5, 36
ventilation 31–3
ventricle, heart 35, 36–7
ventricular fibrillation 95
venules 39–40
vertebrates 101
vertical gene transmission 110–11
vesicle 9, 17, 22
virus 84–89, 102, 111
vital capacity 32–3
vitamins 73
water
  in diet 73
  potential 18–9, 46
  role in living organisms 52–3
  transport in plants 45–50
Watson, James 65
whooping cough 90–1
wine 80, 82
Worldwide Fund for Nature
  (WWF) 116

xerophytes 50
X–ray diffraction 65
xylem 24, 26, 44, 45–7, 51

yeast 22, 80
yoghurt 81, 82–3

zonation 98
zoos 114
zygote 23